AF750838

Energy, Environment, and Sustainability

Series Editor

Avinash Kumar Agarwal, Department of Mechanical Engineering, Indian Institute of Technology Kanpur, Kanpur, India

AIMS AND SCOPE

This books series publishes cutting edge monographs and professional books focused on all aspects of energy and environmental sustainability, especially as it relates to energy concerns. The Series is published in partnership with the International Society for Energy, Environment, and Sustainability. The books in these series are edited or authored by top researchers and professional across the globe. The series aims at publishing state-of-the-art research and development in areas including, but not limited to:

- Renewable Energy
- Alternative Fuels
- Engines and Locomotives
- Combustion and Propulsion
- Fossil Fuels
- Carbon Capture
- Control and Automation for Energy
- Environmental Pollution
- Waste Management
- Transportation Sustainability

Review Process

The proposal for each volume is reviewed by the main editor and/or the advisory board. The chapters in each volume are individually reviewed single blind by expert reviewers (at least four reviews per chapter) and the main editor.

Ethics Statement for this series can be found in the Springer standard guidelines here https://www.springer.com/us/authors-editors/journal-author/journal-author-helpdesk/before-you-start/before-you-start/1330#c14214

Hardikk Valera · Avinash Kumar Agarwal
Editors

Ammonia and Low Carbon Combustion Engine Technologies for De-Fossilizing Transport Sector

Springer

Editors
Hardikk Valera
Engine Research Laboratory
Department of Mechanical Engineering
Indian Institute of Technology Kanpur
Kanpur, Uttar Pradesh, India

Avinash Kumar Agarwal
Engine Research Laboratory
Department of Mechanical Engineering
Indian Institute of Technology Kanpur
Kanpur, Uttar Pradesh, India

ISSN 2522-8366 ISSN 2522-8374 (electronic)
Energy, Environment, and Sustainability
ISBN 978-981-96-6627-0 ISBN 978-981-96-6628-7 (eBook)
https://doi.org/10.1007/978-981-96-6628-7

© The Editor(s) (if applicable) and The Author(s), under exclusive license to Springer Nature Singapore Pte Ltd. 2026

This work is subject to copyright. All rights are solely and exclusively licensed by the Publisher, whether the whole or part of the material is concerned, specifically the rights of translation, reprinting, reuse of illustrations, recitation, broadcasting, reproduction on microfilms or in any other physical way, and transmission or information storage and retrieval, electronic adaptation, computer software, or by similar or dissimilar methodology now known or hereafter developed.
The use of general descriptive names, registered names, trademarks, service marks, etc. in this publication does not imply, even in the absence of a specific statement, that such names are exempt from the relevant protective laws and regulations and therefore free for general use.
The publisher, the authors and the editors are safe to assume that the advice and information in this book are believed to be true and accurate at the date of publication. Neither the publisher nor the authors or the editors give a warranty, expressed or implied, with respect to the material contained herein or for any errors or omissions that may have been made. The publisher remains neutral with regard to jurisdictional claims in published maps and institutional affiliations.

This Springer imprint is published by the registered company Springer Nature Singapore Pte Ltd.
The registered company address is: 152 Beach Road, #21-01/04 Gateway East, Singapore 189721, Singapore

If disposing of this product, please recycle the paper.

Preface

The use of non-carbon-based fuel is in focus in various sectors for various countries to lower greenhouse gas emissions. Ammonia is getting attention from policymakers and researchers as it doesn't contain any carbon molecules in its structure. Along with ammonia, other low-carbon combustion technologies are also in focus to lower the emissions from the transport sector.

The International Society for Energy, Environment and Sustainability (ISEES) was founded at the Indian Institute of Technology Kanpur (IIT Kanpur), India, in January 2014 to spread knowledge and awareness and to catalyze research activities in the fields of Energy, Environment, Sustainability, and Combustion. Society's goal is to contribute to the development of clean, affordable, and secure energy resources and a sustainable environment for society, spread knowledge in the areas mentioned above, and create awareness about the environmental challenges the world is facing today. The unique way adopted by ISEES was to break the conventional silos of specializations (engineering, science, environment, agriculture, biotechnology, materials, fuels, etc.) to tackle the problems related to energy, environment, and sustainability holistically. This is evident by the participation of experts from all fields to resolve these issues. The ISEES is involved in various activities, such as conducting workshops, seminars, and conferences in the domains of its interests. The society also recognizes the outstanding works of young scientists, professionals, and engineers for their contributions by conferring awards in various categories.

The Eighth International Conference on 'Sustainable Energy and Environmental Challenges' (VIII-SEEC) was organized under the auspices of ISEES from December 04 to 06, 2023, at the Malaviya National Institute of Technology Jaipur, India. This conference provided a platform for discussions between eminent scientists and engineers from several countries, including India, the USA, Brazil, Canada, France, Greece, and Africa. At this conference, eminent international speakers presented their views on energy, combustion, emissions, and alternative energy resources for sustainable development and a cleaner environment. The conference presented two high-voltage plenary talks by Prof. G. D. Yadav, Institute of Chemical Technology, Mumbai, and Prof. Norma A. Alcantar, University of South Florida, USA.

The conference included 21 technical sessions on energy and environmental sustainability topics, including two plenary talks, 23 keynote talks, 35 invited talks from prominent scientists, and 95 contributed talks by students and researchers. The conference included technical sessions on Alternative Fuels for Sustainable Transport, Advanced Engines and Fuels, Sustainable Technologies, New Energy, Biomass and Gasification, Advanced Diagnostics and Optical Techniques, Policies for Environmental Sustainability, Solid Waste Remediation, Wastewater Treatment, Air Pollution and Control, Natural Treatment System, Pollution Mitigation, Environmental Sustainability: Challenges and Priorities, Energy and Environment, Sustainable Processing of Biomass, Biofuels and Biorefineries, Environmental Bioengineering, Microbial Processes, Bioremediation, Human Health and Environmental Sustainability, and several assorted topics.

250+ participants and speakers attended the conference, where six ISEES books published by Springer, Singapore, under a dedicated series, *Energy, Environment and Sustainability*, were released. This conference laid the roadmap for technology development, opportunities, and challenges in the Energy, Environment, and Sustainability domain. These topics are relevant for the country and the world in the present context. We acknowledge the support from various funding agencies and organizations for the successful conduct of the VIII-SEEC, where the idea of these books germinated. We would therefore gratefully like to acknowledge Malaviya National Institute of Technology Jaipur (Special thanks to Dr. Nikhil Sharma); SERB, Government of India; Department of Scientific and Industrial Research (DSIR) (Special thanks to Dr. Vipin Shukla), Sponsors (Dynomerk Controls India Pvt. Ltd., Rebound Envirotech Pvt. Ltd., BlueDrop Enviro Pvt. Ltd.), and our publishing partner Springer (Special thanks to Ms. Swati Mehershi).

The editors would like to express their sincere gratitude to many authors worldwide for submitting their high-quality work on time and revising it appropriately at short notice. We want to express our special gratitude to our prolific reviewers: Dr. P. A. Lakshminarayanan, Dr. Daanish Tyrewala, Dr. Snehasis Panigrahy, Dr. Niranjan Miganakallu, and several potential researchers who reviewed chapters of this monograph and provided their valuable suggestions to improve them.

This book discusses the various aspects of ammonia and low-carbon combustion engine technologies for de-fossilizing the transport sector. This book comprehensively reviews ammonia as an alternative fuel for internal combustion engines. This book includes a few chapters on fundamental aspects of ammonia combustion while providing information about kinetic mechanisms. This book also contains a few chapters on ammonia engine technologies. A few chapters include recent results and are focused on current trends in ammonia-fueled internal combustion engines. This book is a valuable reference for those seeking to stay at the forefront of these evolving fields. We hope the book greatly interests the professionals, researchers, and students working on Ammonia and low-carbon engine combustion.

Kanpur, India

Hardikk Valera
Avinash Kumar Agarwal

Contents

About the Editors

Mr. Hardikk Valera is a Ph.D. scholar at the Indian Institute of Technology Kanpur (IITK). He has completed his M.Tech. from the National Institute of Technology Jalandhar (India) and B.Tech. from Ganpat University. His research interests include methanol-fueled SI engines, methanol-fueled CI engines, spray characterization, and emission control from engines. He has published 15 peer-reviewed international journal papers and two international conference papers, seven edited books, and 19 chapters, with 230+ Scopus and 350+ Google Scholar citations.

Prof. Avinash Kumar Agarwal is Director at IIT Jodhpur. He obtained his Undergraduate Degree in Mechanical Engineering (1994) from Malviya Regional Engineering College, Jaipur, and his M.Tech. (Energy, 1996) and Ph.D. (Energy, 1999) from the Indian Institute of Technology (IIT) Delhi. After his Post-doctoral Fellowship (1999–2001) at the ERC, UW, Madison, USA, he returned to India in 2001 and joined IIT Kanpur. Professor Agarwal has established a state-of-the-art "Engine Research Laboratory" at IIT Kanpur. Professor Agarwal is interested in researching IC engines, combustion, conventional and alternative fuels, optical diagnostics, laser ignition, HCCI, particulates and emission control, and large bore engines. Professor Agarwal has published over 550+ peer-reviewed international journal and conference papers, 65 edited books, and 129 book chapters, attracting 18300+ Scopus and 27400+ Google Scholar citations.

For his outstanding contributions, Prof. Agarwal is conferred upon Sir J. C. Bose National Fellowship (2019) by SERB, Prestigious Shanti Swarup Bhatnagar Prize (2016) in Engineering Sciences, Rajib Goyal Prize in Physical Sciences (2015); NASI-Reliance Industries Platinum Jubilee Award (2012); INAE Silver Jubilee Young Engineer Award (2012); INSA Young Scientist Award (2007); UICT Young Scientist Award (2007); INAE Young Engineer Award (2005); Devendra Shukla Research Fellowship (2009–2012), Poonam and Prabhu Goyal Endowed Chair Professorship (2013–2016), SBI Endowed Chair Professorship (2018–2021, 2022–2025) at IIT Kanpur. He is an elected Fellow of TWAS, SAE, ASME, INAE, ISEES, RSC, NASI, WSSET, AAAS and Combustion Institute USA.

Part I
Ammonia and Low Carbon Engine Combustion

Chapter 1
Introduction to Ammonia and Low-Carbon Combustion Engine Technologies for De-Fossilizing the Transport Sector

Hardikk Valera and Avinash Kumar Agarwal

Abstract Globally, the race towards Net-Zero emissions has started. The country focuses on eliminating carbon dioxide and carbon monoxide emissions from the atmosphere. Currently, the dominating energy source is fossil-based, but the Net Zero vision warrants using an energy source that doesn't contain carbon in its molecule. The transport sector is under stress as regulatory bodies put immense pressure on lowering tail-pipe emissions. Also, the power industries are looking for a fuel that uses non-carbonaceous radicals for combustion. Ammonia is the future fuel for decarbonizing the transport and power sectors. Also, low-carbon combustion technologies are needed to meet Net-Zero targets. The engine manufacturers are looking to retrofit the existing internal combustion engines to enable Ammonia combustion. The high auto-ignition temperature and slower chemical kinetics cause difficulty in initiating the ammonia combustion. Thus, the ignition or combustion improver is needed to begin the combustion. Although ammonia use in the transport and power sector is nascent, scientists are working on overcoming the challenges. This book covers the current challenges and state-of-the-art reviews of ammonia-fueled internal combustion engines, critical equipment and system-based risk factors of an ammonia-fueled, safety aspects of ammonia utilization, pre-chamber ignition system for lean combustion in a spark ignition engine, and recent optical techniques for soot characterization in fuel spray flames.

Keywords Net Zero · Ammonia · Combustion · Kinetics · Safety equipment

The global imperative for decarbonization has spurred significant interest in alternative fuels for internal combustion engines. Ammonia emerges as a promising candidate for decarbonizing transport. Carbon-free nature, low cost subjected to production

---

H. Valera · A. K. Agarwal (✉)
Engine Research Laboratory, Department of Mechanical Engineering, Indian Institute of Technology Kanpur, Kanpur, India
e-mail: akag@iitk.ac.in

© The Author(s), under exclusive license to Springer Nature Singapore Pte Ltd. 2026
H. Valera and A. K. Agarwal (eds.), *Ammonia and Low Carbon Combustion Engine Technologies for De-Fossilizing Transport Sector*, Energy, Environment, and Sustainability, https://doi.org/10.1007/978-981-96-6628-7_1

route, existing supply infrastructure, etc., highlight features that make ammonia a superior fuel. However, like any other fuel, it also has certain disadvantages that must be addressed. High nitrogen oxide emissions and unburnt Ammonia are the two broad challenges for Ammonia as a fuel that can be overcome by regulating the combustion process. In parallel, low-carbon combustion technologies are also getting attention for decarbonizing transport.

The first section of this book is Ammonia and Low-Carbon Engine Combustion. The first chapter, Introduction to Ammonia and Low-Carbon Combustion Engine Technologies for the De-Fossilizing Transport Sector, introduces this monograph. The second chapter comprehensively reviews ammonia as an alternative fuel for internal combustion engines. This chapter explores the potential of ammonia as a carbon-free fuel for internal combustion engines, focusing on the promising dual-fuel combustion strategies. While ammonia offers significant environmental benefits, including reduced greenhouse gas emissions and improved air quality, its implementation faces challenges such as low flame propagation speed, high autoignition temperature, increased nitrogen oxides, and unburned ammonia emissions. The third chapter is about designing a pre-chamber jet ignition system for the lean combustion of fuels. This chapter reviews the state of the art of developing different parameters of the prechamber. The physical phenomenon inside the prechamber and its importance in efficient design have been discussed. The effect of different prechamber parameters on physical phenomena in the prechamber, combustion, emissions, and performance of an engine, such as flame front propagation, charge motion in the prechamber, jet ejection, and scavenging processes, in-cylinder pressure, heat release rate, are discussed. This chapter also discusses constant-volume combustion chambers, rapid compression machines, and engine studies conducted for the design optimization of pre-chambers. The fourth chapter is Fuel Properties and Suitability for Internal Combustion Engines: Exploring Conventional, Alternative, and Bio-Blended Fuels. This chapter outlines bio-blends, including biodiesel, bioethanol, and biogas, highlighting their properties like renewability, lower greenhouse gas emissions, and compatibility with existing engine technologies. This chapter underscores the potential of these bio-blends to enhance energy security and environmental sustainability, demonstrating their growing importance in the future of engine technology.

The second section of this book is Fundamentals of Ammonia Combustion, which contains three chapters. There is a need to understand the fundamentals of ammonia combustion and fine-tune the combustion process for widespread use. The first chapter is Insights into Fundamentals of Ammonia Combustion. This chapter will help understand the details of the ammonia combustion/oxidation process, which will help achieve greenhouse gas emission targets soon. The second chapter is decoding the challenge of ammonia combustion using a hybrid approach, considering computational fluid dynamics and artificial intelligence. Considering such a mechanism in a computational fluid dynamics framework, a high-fidelity model is first constructed to understand the behavior of several species under the combustion process. The third chapter is Kinetic Mechanism Development and CFD Study of Combustion of Ammonia Blends in a Micro-Gas Turbine. This study focused on two main objectives: (a) developing a semi-detailed $NH_3/CH_4/H_2/n\text{-}C_{12}H_{26}$ chemical mechanism involving 242 species and 1769 reactions, including sub-mechanisms for nitrogen oxides and soot formation, to analyze the combustion and emission behavior of ammonia-fuel blends in engines; and (b) applying this mechanism to study the combustion properties of a simplified micro-gas turbine using a steady flamelet model. The mechanism was validated using experimental and numerical data from the literature.

The book's last section is Ammonia Engine Technologies, with three chapters. The first chapter is a state-of-the-art review of ammonia-based diesel engines. This chapter explores ammonia's combustion characteristics, energy density, and storage requirements. Technological advancements, such as dual-fuel systems and catalytic converters, are evaluated for optimizing ammonia usage in diesel engines. The review also addresses economic and safety considerations, highlighting the feasibility of Ammonia as a sustainable alternative to conventional diesel fuel in achieving cleaner transportation solutions. The second chapter is A Comprehensive Review on the Emission Characteristics of Ammonia-Powered Engines. Its high-octane number permits the advancing of spark timing, offering an excellent knock resistance property. Also, the ease of production, storage, and transportation makes it a viable option. The complete combustion of ammonia produces only nitrogen and water, but due to the oxidation of N_2 at elevated temperatures, it produces thermal NO_X. The challenges, such as high N_2O, unburned ammonia, and hydrogen cyanide emissions due to fuel-bound nitrogen, must be addressed for effective ammonia combustion in internal combustion engines. This chapter provides a detailed review of the emission characteristics of ammonia-powered engines. The third chapter is Identification of Critical Equipment and System-Based Risk Factors of an Ammonia-Fueled Marine Diesel Engine. Utilizing ammonia could help achieve the IMO's 2050 net-zero target. However, ammonia's unique properties must be carefully managed. As a toxic and corrosive substance, ammonia poses significant risks to the marine ecosystem if spilled. Additionally, being a flammable gas, it carries flash, explosion, and fire risks. If stored at −33 °C as a cold liquid, Ammonia also presents a cold-burn hazard in the event of a fuel system leak. This chapter identifies critical equipment and system-based risk factors associated with ammonia fuel supply systems and ammonia-fueled marine diesel engines. It highlights key equipment and potential risk factors, including toxicity, corrosiveness, stress corrosion, low-temperature embrittlement, and flammability.

This monograph is divided into three different sections: (I) Ammonia and Low-Carbon Engine Combustion, (II) Fundamentals of Ammonia Combustion, and (III) Ammonia Engine Technologies. Specific chapters covered in the manuscript include:

- Introduction to Ammonia and Low-Carbon Combustion Engine Technologies for De-Fossilizing the Transport Sector
- Comprehensive Review of Ammonia as an Alternative Fuel for Internal Combustion Engines
- Design of Prechamber Jet Ignition System for Lean Combustion of Fuels
- Fuel Properties and Suitability for Internal Combustion Engines: Exploring Conventional, Alternative, and Bio-Blended Fuels

- Insights into Fundamentals of Ammonia Combustion
- Decoding the Challenge of Ammonia Combustion: A Hybrid Approach considering CFD and AI
- Kinetic Mechanism Development and CFD Study of Combustion of Ammonia Blends in a Micro-Gas Turbine
- State-of-the-Art Review on Ammonia-Based Diesel Engines
- A Comprehensive Review on the Emission Characteristics of Ammonia-Powered Engines
- Identification of Critical Equipment and System-based Risk Factors of an Ammonia-Fueled Marine Diesel Engine.

Chapter 2
Comprehensive Review of Ammonia as an Alternative Fuel for Internal Combustion Engines

Ayat Gharehghani, Mohammadreza Azizi, and Jabraeil Ahbabi Saray

Abstract The global imperative for decarbonization has spurred significant interest in alternative fuels for internal combustion engines. Ammonia emerges as a promising candidate due to its complete combustion producing only nitrogen and water as byproducts. This review investigates the potential of ammonia as a carbon-free fuel for internal combustion engines (ICEs), focusing on the promising dual-fuel combustion strategies. While ammonia offers significant environmental benefits, including reduced greenhouse gas emissions and improved air quality, its implementation faces challenges such as low flame propagation speed, high autoignition temperature, and increased NOx, N_2O and unburned ammonia emissions. The review explores current research efforts to address these issues, including the use of combustion promoters and optimized engine parameters. Despite these technical obstacles, ammonia's potential to contribute to a sustainable future is evident, and ongoing research and development are crucial to unlock its full potential as a clean and efficient fuel for the transportation sector.

Keywords Ammonia · Internal combustion engine · Greenhouse gas emissions · Carbon-free fuel · Sustainability

2.1 Introduction

The extensive use of fossil fuels has precipitated a critical environmental issue: greenhouse gas emissions. Decarbonizing internal combustion engines is pivotal in alleviating these impacts, with ammonia emerging as a promising, sustainable, carbon-free fuel across multiple sectors. According to Fig. 2.1, ammonia's use as a transportation fuel can be traced back to the early 19th century (Cardoso et al. 2021). Despite increasing attention over the past decade, ammonia combustion remains

---

A. Gharehghani (✉) · M. Azizi · J. A. Saray
School of Mechanical Engineering, Iran University of Science and Technology, Tehran, Iran
e-mail: ayat_gharehghani@iust.ac.ir

© The Author(s), under exclusive license to Springer Nature Singapore Pte Ltd. 2026

H. Valera and A. K. Agarwal (eds.), *Ammonia and Low Carbon Combustion Engine Technologies for De-Fossilizing Transport Sector*, Energy, Environment, and Sustainability, https://doi.org/10.1007/978-981-96-6628-7_2

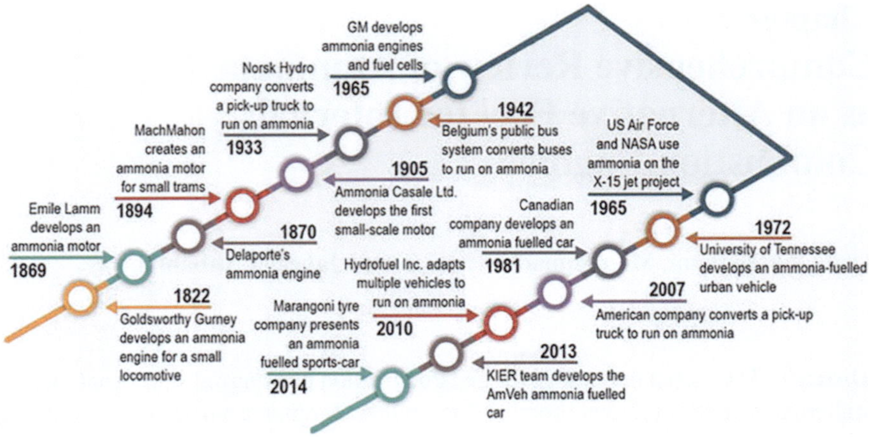

Fig. 2.1 The advancements in using ammonia as a fuel for internal combustion engines (Cardoso et al. 2021)

under-researched and faces implementation challenges, highlighting the necessity for comprehensive studies in this domain (Blasio et al. 2022).

Studies on ammonia engines have a historical backdrop dating to 1904, evolving through three distinct phases. During World War II, restricted fossil fuel access spurred significant research into ammonia engines. In 1943, Belgium pioneered ammonia's use in internal combustion engines amid wartime gasoline and diesel constraints. Adapting six buses to operate on anhydrous ammonia and coal gas, these engines successfully ran in spark ignition (SI) mode, showcasing early promise for ammonia as a viable engine fuel (Bjørgen et al. 2024). Following the war, interest in ammonia engines diminished with the resurgence of petroleum availability. However, energy crises in the 1960s and 1970s reignited interest despite ammonia's high costs, challenging its widespread adoption (Zhou 2023). In recent years, urgent imperatives to reduce greenhouse gas emissions have rekindled interest in ammonia engines as a potential clean energy source (Zhou 2023).

Ammonia (NH3), a colorless gas with a pungent odor at ambient conditions, is an essential compound comprised of nitrogen and hydrogen. In contemporary industry, ammonia is vital, especially in agriculture, where it is a primary ingredient in chemical fertilizers. Furthermore, ammonia shows significant potential as a clean energy source for the future (Dimitriou and Javaid 2020).

Ammonia exhibits several crucial physical properties. It has a boiling point of −33.3 °C (240 K) and a freezing point of −77.8 °C (195.5 K), making it a gas at standard temperature and pressure. At standard temperature, ammonia has a density of 0.73 g/cm^3. Additionally, ammonia has a relatively high autoignition temperature of 651°C (924 K), meaning it requires substantial heat to ignite without an external flame. These properties are critical for its storage, handling, and application in various industrial contexts, including refrigeration, fertilizers, and as a precursor to numerous nitrogen-containing compounds. Ammonia inhalation can cause significant damage to the respiratory system. Fortunately, its pungent and irritating odor makes it readily detectable. Even at low concentrations (around 0.6–53 ppm), which are well below the occupational exposure limits. The "immediately dangerous to life or health (IDLH)" concentration for ammonia is 300 ppm. This means that exposure to such concentrations, even for a short period, can result in severe health consequences, including death (Giddey et al. 2013). Ammonia's lower density compared to air causes it to dissipate quickly in the atmosphere when released, minimizing the risk of extended exposure in open areas (Dimitriou and Javaid 2020). While natural gas is also lighter than air, its dispersion rate can vary depending on specific conditions, such as wind speed and atmospheric stability.

Ammonia stands out as an exceptional fuel for combustion applications, offering a compelling array of benefits. Ammonia boasts a remarkable volumetric energy density compared to other energy carriers like hydrogen, facilitating simpler storage and transportation. It can be decomposed into hydrogen, making it a viable fuel for fuel cells and hydrogen-powered engines. The existing infrastructure for ammonia transport, including pipelines, roads, and railways, is well-established in many countries, making its storage and movement safer and more cost-effective than hydrogen. Additionally, the Haber-Bosch process, widely employed in the nitrogen fixation industry, provides a cost-effective and well-established method for ammonia production. Furthermore, ammonia has the potential to replace fossil fuels, thereby contributing to a decline in greenhouse gas emissions (Lee et al. 2010; Ahbabi Saray et al. 2024). Ammonia offers several advantages, including high energy density, zero carbon emissions, existing infrastructure, safety, and affordability. These benefits make it a promising option for a cleaner and more sustainable energy future (Yapicioglu and Dincer 2018). Industry experts and analysts anticipate a significant growth trajectory for the ammonia market in the coming decade. According to projections by Yapicioglu and Dincer (2018), the global ammonia market is poised for a 3% growth rate over the next ten years. This forecasted growth signifies a pivotal turning point in ammonia's journey toward becoming a mainstream fuel source. The widespread adoption of ammonia as a fuel holds immense potential for environmental benefits. By replacing conventional fossil fuels, ammonia combustion can substantially reduce greenhouse gas emissions, mitigating the harmful effects of climate change. Additionally, ammonia production can be powered by renewable energy sources, further minimizing its environmental footprint. The transition to ammonia-powered technologies will require advancements in engine design and optimization to address the current challenges associated with ammonia combustion. Research efforts are focused on developing efficient and reliable ammonia combustion systems that can meet the performance and emissions standards of modern engines. As research and development progress, ammonia is poised to play an increasingly significant role in the global energy landscape. Ammonia's inherent advantages, combined with the rising demand for sustainable energy, make it a promising candidate to transform the way we power our transportation, industries, and homes.

2.1.1 Technical Challenges and Potential Solutions

The strategies and key parameters for using ammonia as fuel in internal combustion engines are illustrated in Fig. 2.2. Ammonia's combustion characteristics pose significant challenges for direct engine use. Researchers are investigating various combustion promoters, tailored to different engine types. Figure 2.3 demonstrates how varying ammonia concentrations affect combustion efficiency and emissions in various fuel mixtures. For spark-ignition (SI) engines, hydrogen is a prevalent choice, either pre-stored or generated from ammonia itself. Other studies have explored gasoline, alkanes or alcohols as promoters in SI engines. Conversely, compression-ignition (CI) engines require highly reactive promoters like diesel or dimethyl ether (DME) to effectively ignite the ammonia (Zhu and Shu 2023). Optimizing key engine parameters is crucial; compared to traditional fuels, ammonia's slower burning rate often necessitates earlier ignition timing. High-performance ignition systems can also enhance combustion stability. Designing and optimizing the injection system presents a significant challenge, as ammonia can be injected as a gas or liquid, either through ports or directly into the cylinder. While adding promoters increases flexibility, it also adds complexity (Kumar et al. 2024). As shown in Fig. 2.4, the fuel supply system in ammonia-blended dual fuel engines is complex and the promoter for blending requires a separate space and a separate storage container for storage. Other factors, such as compression ratio, intake air pressure, and temperature, significantly impact engine performance. Although some research has explored these aspects, more systematic investigations are needed to fully understand and optimize ammonia combustion in engines (Zhu and Shu 2023).

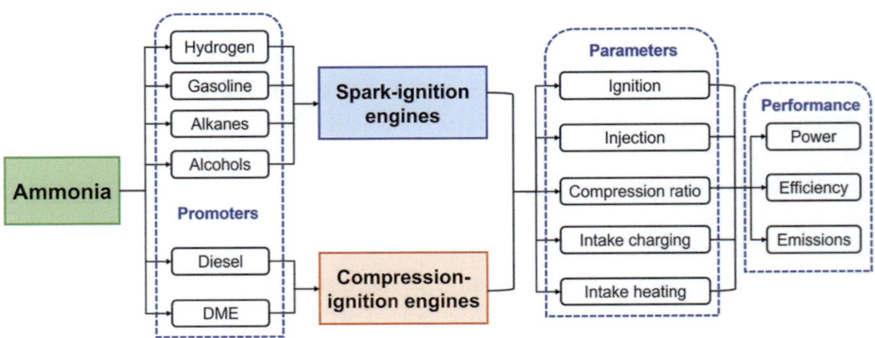

Fig. 2.2 Strategies and key parameters for using ammonia as fuel in internal combustion engines (Zhu and Shu 2023)

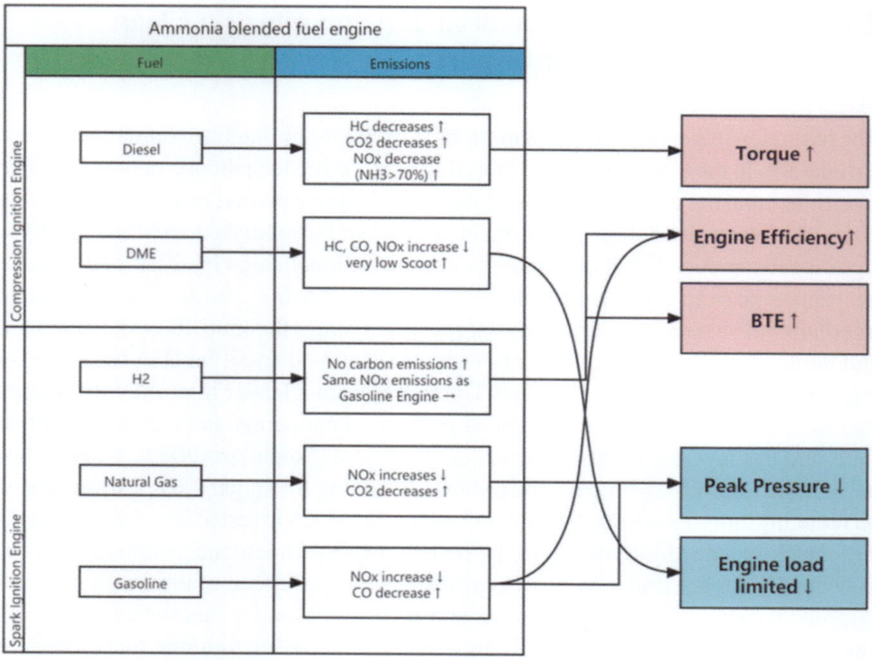

Fig. 2.3 Summary of the effects of different fuel mixtures (Xu et al. 2022)

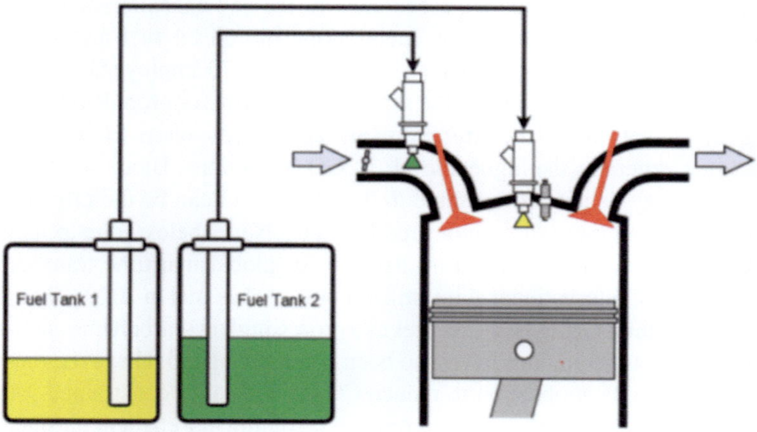

Fig. 2.4 Schematic diagram of fuel supply system for ammonia-blended dual fuel engine (Xu et al. 2022)

2.2 The Necessity and Potential of Ammonia as a Clean Fuel in Maritime Transport

The reduction of greenhouse gas emissions (GHGs) stemming from human activities, particularly in the transportation sector, stands as a critical challenge of our time. The maritime industry, a cornerstone of global trade, plays a pivotal role in this realm. In 2018, the International Maritime Organization (IMO) established ambitious goals to significantly reduce GHG emissions from international ships by 2050 and achieve net-zero emissions by the end of the century. Given the 20–30 year lifespan of international vessels, realizing these targets necessitates the immediate development and deployment of alternative solutions to conventional fossil fuels in the shipping industry (Zhou 2023). The European Union has been a leader in promoting cleaner maritime transport. It has implemented policies to encourage the use of alternative fuels and has invested in infrastructure development (Soldatkina 2023). Norway has been at the forefront of electric and hybrid propulsion for ships. It has a strong focus on reducing emissions from its maritime sector. Japan has invested heavily in research and development of alternative fuels, particularly hydrogen and ammonia. It aims to become a global leader in clean marine technology. As one of the world's largest shipping nations, China is investing in alternative fuels and technologies to meet its environmental goals. The United States has been exploring options for alternative fuels in marine transportation, including biofuels and LNG (Dimitriou and Javaid 2020; Tu et al. 2024; Lee et al. 2024).

Table 2.1 presents a comparative overview of the key characteristics of various low-carbon and zero-carbon marine fuels, alongside diesel fuel and lithium-ion battery with 260 Wh/kg energy density (i.e. Panasonic 2170 employed in Tesla Model 3). Diesel engines have dominated the market as the primary propulsion system for ocean-going vessels due to their durability and cost-effectiveness. However, the use of diesel fuel generates the highest levels of CO_2 emissions. Biodiesel possesses a volumetric energy density comparable to diesel fuel and can be directly employed in marine diesel engines with minor modifications. Nevertheless, due to its limited production capacity and restricted availability for global maritime transportation, less than 0.02% of international ships utilized biodiesel as fuel in 2020 (Zhou 2023).

Liquefied Natural Gas (LNG) emerges as a promising low-carbon marine fuel, and corresponding marine engines have also been commercialized. However, the potential for expanding its application to reduce CO_2 emissions is constrained. Methane slip, particularly at low engine loads, remains a significant challenge. Since LNG must be stored at a cryogenic temperature of −162 °C, the cost of an LNG fuel tank is approximately 43.9 times that of a diesel fuel tank. Consequently, while 581 international ships were equipped with dual-fuel LNG/diesel engines in 2020, approximately 70% of them were LNG carriers (Zhou 2023). Methanol stands out as a promising alternative to diesel fuel. Remaining in a liquid state under ambient conditions, it offers several advantages. Its energy density significantly exceeds that of liquid hydrogen, facilitating easier storage and transportation. By utilizing green methanol, CO_2 emissions can be entirely eliminated during the combustion process. However,

Table 2.1 A comparative overview of the key characteristics of various low-carbon and zero-carbon marine fuels, alongside diesel fuel and lithium-ion batteries

Fuel (Percent for international. Ship prime power)	Fuel density (kg/m^3)	Lower heating value (MJ/kg)	Energy density (MJ/L)	Fuel and tank volume per heat (Diesel = 1)	Storage capex cost (Diesel = 1)	CO_2 emissions per unit of heat (Diesel = 1)
Diesel (99.38%)	840	42.5	35.7	1	1	1
Biodiesel (0.02%)	880	38.3	33.7	1.1	1.1	0
Methanol (0.01%)	800	19.9	15.9	2.2	2.2	0.9
LNG (0.59%)	480	48	23	2.5	43.9	0.74
Liquid ammonia	603	18.8	11.3	3.1	10.9	0
Liquid hydrogen	71	120	8.5	6.7	137.7	0
Lithium battery	1250	0.9	1.1	52.7	14511	0

green methanol production necessitates advanced carbon capture, utilization, and storage (CCUS) technologies, which are currently in the early stages of development and relatively expensive (Zhou 2023; Ghandehariun et al. 2024). Hydrogen is touted as a sustainable and carbon-free fuel. During combustion, hydrogen releases no direct carbon dioxide emissions and boasts a high volumetric energy density. However, significant challenges hinder the widespread adoption of hydrogen in the marine transportation industry. Storing and transporting hydrogen is both challenging and costly due to its volatile nature. Liquid hydrogen must be maintained at an extremely low temperature of –253 °C, necessitating specialized tanks and equipment. Additionally, the volumetric energy density of liquid hydrogen is considerably lower than that of fossil fuels, implying that larger tank volumes and storage spaces are required to cover similar distances (Zhou 2023). Batteries, as electrical energy storage devices, offer unique advantages. They operate emission-free and can be charged using renewable sources like solar or wind energy. Nevertheless, substantial challenges arise when considering large-scale battery usage for ships. The volumetric energy density of current batteries falls significantly short of that of fossil fuels. Consequently, powering ocean-going vessels over long distances would demand massive, heavy batteries that occupy considerable space, as depicted in Fig. 2.5, and impact the ship's cargo-carrying capacity. Moreover, as shown in Fig. 2.5, the cost of batteries remains extremely high, and their lifespan is limited. Additionally, the disposal of spent batteries poses a significant environmental concern, and the

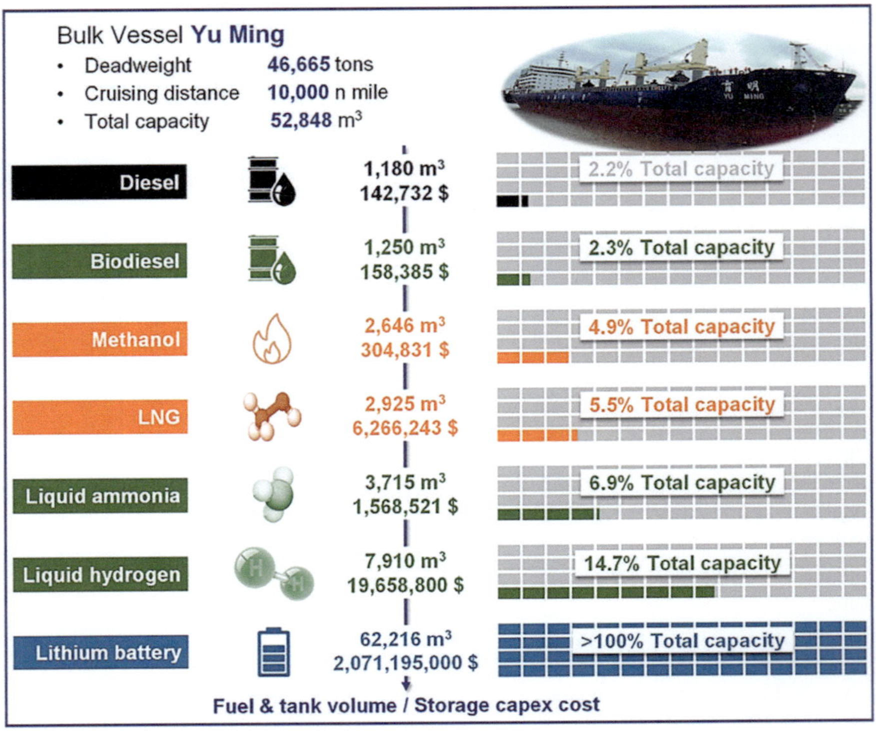

Fig. 2.5 Volume of fuel and tank required for a 10,000 nautical mile (nm) sea voyage using various low-carbon and carbon-free marine fuels (diesel fuel and lithium-ion batteries are also included for comparison) (Zhou 2023)

recycling infrastructure for batteries is still in its early stages. Given these challenges, employing batteries as the sole propulsion source for large ships is currently economically unjustified (Zhou 2023).

2.2.1 Ammonia: A Promising Option as a Carbon-Free Marine Fuel

Exploring low-carbon and carbon-free alternatives to fossil fuels in the maritime transportation industry, we turn our attention to ammonia. This promising fuel offers significant advantages in storage, mass production, and sustainability, making it an attractive choice for the future. Ammonia remains liquid at room temperature under a pressure of approximately 9 bar, reducing storage and transportation challenges compared to alternatives like liquid hydrogen, which requires extremely low temperatures and specialized tanks. Initial investment and storage costs for liquid ammonia tanks are estimated to be between 3.6% and 7.9% of those for liquid hydrogen tanks

(Sagel et al. 2022). The infrastructure for ammonia production and transportation is well-established, primarily through the Haber-Bosch process, which combines nitrogen separated from air with hydrogen (Cha et al. 2021). Various projects worldwide are currently underway to produce green ammonia on a large scale. For instance, Yara, an international ammonia production giant, has commissioned a green ammonia plant in Australia (Valera-Medina et al. 2018). Similarly, Canada plans to construct a green ammonia production facility powered by hydropower in Quebec by 2025. This hub will have a monthly capacity of 15,000 tons of green ammonia, catering to the maritime sector. Ammonia can be generated by combining hydrogen derived from renewable energy sources with nitrogen separated from air, resulting in "green ammonia," positioning it as a sustainable and environmentally friendly option. Given ammonia's numerous advantages in storage, mass production, and sustainability, it has garnered significant attention as a carbon-free fuel for marine engines. Technological advancements and investments in green ammonia production projects amplify its potential to play a pivotal role in a sustainable future for the maritime transportation industry (Zhou 2023).

There are two main types of engines that can be used with ammonia: two-stroke engines and four-stroke engines. Two-stroke engines are typically used in smaller ships, while four-stroke engines are used in larger ships.

Several fleets are already testing ammonia as a marine fuel. For example, the Norwegian shipping company Wilhelmsen New Energy is testing ammonia in a 2500-ton bulk carrier. The Japanese shipping company Mitsui O.S.K. Lines is also testing ammonia in a 2500-ton bulk carrier. The size of the engines that can be used with ammonia varies. However, most engines that are being tested with ammonia are large engines, typically over 10,000 horsepower (Micco et al. 2024).

There are several challenges that need to be overcome before ammonia can be widely used as a marine fuel. These challenges include the need to develop a safe and efficient way to store and transport ammonia, as well as the need to develop new engine technologies that can use ammonia as a fuel. Despite these challenges, ammonia is a promising option as a carbon-free marine fuel. If these challenges can be overcome, ammonia could help to reduce greenhouse gas emissions from the shipping industry. Here is a table that summarizes the fleets that are testing ammonia as a marine fuel.

In addition to the fleets listed in Table 2.2, there are also several research projects that are investigating the use of ammonia as a marine fuel. These projects are studying the technical feasibility of using ammonia as a marine fuel, as well as the economic and environmental benefits of using ammonia as a marine fuel.

2.3 Dual-Fuel Ammonia Combustion in CI Engines

The strategies and key parameters for using ammonia as fuel in internal combustion engines are illustrated in Fig. 2.2. Ammonia's combustion characteristics pose significant challenges for direct engine use. Researchers are investigating various

Table 2.2 Ammonia-powered marine vessels: fleet trials and specifications

Fleet	Ship type	Engine type	Size
Wilhelmsen new energy	Bulk carrier	Two-stroke	2500 tons
Mitsui O.S.K. lines	Bulk carrier	Two-stroke	2500 tons
MISC	Tanker	Four-stroke	160,000 tons
Höegh autoliners	Car carrier	Four-stroke	7500 TEU[1]

combustion promoters, tailored to different engine types. Figure 2.3 demonstrates how varying ammonia concentrations affect combustion efficiency and emissions in various fuel mixtures (Shi 2024).

Diesel engines remain the primary propulsion system for heavy-duty transportation, power generation, and marine vessels due to their high efficiency and ability to handle heavy loads over extended periods. However, their significant pollution, particularly the emission of soot and nitrogen oxides (NOx), poses a major environmental challenge. While advancements in combustion technology and emission control systems have reduced the release of these pollutants (Musculus et al. 2013; Senecal and Leach 2019), the inherent nature of diesel fuel, leading to CO_2 and greenhouse gas emissions, continues to hinder efforts to lower carbon emissions (Yousefi et al. 2022). Alternative solutions, such as utilizing carbon-free fuels, are being explored to minimize diesel's environmental impact while retaining its advantages. Among these, ammonia has emerged as a promising substitute fuel for internal combustion engines, particularly in marine applications (Dimitriou and Javaid 2020; Singh et al. 2021).

Ammonia's combustion characteristics, including slow flame speed, high autoignition temperature, and a narrow flammability range, pose significant challenges for its use in pure ammonia mode in internal combustion engines. In recent decades, considerable research has focused on utilizing ammonia fuel in CI engines. Ammonia's relatively slow flame propagation rate can be effectively addressed by increasing the compression ratio. For instance, in 1967, Pearsall and Garabedian (1968) were unable to achieve successful operation of an internal combustion engine with direct injection of liquid ammonia, even with a compression ratio (CR) of 30:1. The test engine was modified from an AVDS-1790 V-twin engine, and the inlet air temperature was room temperature. Gray Jr. et al. (1967) demonstrated that ammonia-only mode operated successfully at an intake temperature and gas temperature of 422 K and a compression ratio of 35:1. Achieving a compression ratio of 35:1 is essential for CI engine operation with pure ammonia fuel. Therefore, when using ammonia as an alternative fuel in diesel engines, diesel is typically employed as a co-reactant to enhance ammonia combustion (Shi 2024). Dual-fuel combustion, where ammonia is used alongside conventional fuels, presents a viable solution by overcoming the inherent challenges associated with ammonia's poor combustion characteristics. Ammonia's high autoignition temperature, the temperature at which

[1] The **TEU** in the table stands for **Twenty-Foot Equivalent Unit**. It's a standard unit of measure used in the shipping industry to represent the capacity of cargo containers.

it spontaneously ignites, hinders its direct use as a primary fuel. However, in dual-fuel mode, a pilot fuel with a lower autoignition temperature acts as an initiator, triggering the combustion of the ammonia-air mixture (Zhu and Shu 2023).

Diesel and DME are two common pilot fuels employed in dual-fuel ammonia engines. Each offers unique advantages and disadvantages in facilitating ammonia combustion in dual-fuel mode. The choice between diesel and DME as the pilot fuel for dual-fuel ammonia operation depends on several factors, including availability, cost, emission requirements, and engine compatibility. In terms of emissions, DME is a cleaner-burning fuel, producing fewer pollutants compared to diesel, making it a more environmentally friendly option. It also possesses a high cetane number, which can improve combustion efficiency and reduce emissions. Therefore, DME is considered a more environmentally preferable choice for facilitating ammonia combustion in dual-fuel engines (Sorenson 2001). This results in a lower carbon footprint, contributing less to greenhouse gas emissions. Diesel, on the other hand, is a well-established fuel with a mature infrastructure, simplifying its use in dual-fuel engines. It also boasts a high energy density, which can translate into favorable engine performance and efficiency. Additionally, diesel exhibits good lubricating properties, aiding in engine component protection (Cunanan et al. 2021).

2.3.1 Benefits and Challenges of Ammonia/Diesel Dual-Fuel (ADDF) Combustion

The benefits of using ADDF combustion include a gradual reduction in fossil fuel reliance, implementation simplicity, and performance benefits. By substituting a portion of diesel with ammonia, pollutant emissions and crude oil dependency can be progressively decreased. Existing diesel engines can be modified with relatively minor changes to accommodate ADDF operation. Additionally, ammonia can enhance combustion efficiency and reduce overall fuel consumption. ADDF combustion represents a significant step towards developing cleaner and more sustainable transportation. Ongoing research and advancements in this field hold promise for substantial fossil fuel substitution with ammonia in the future. Ammonia fuel holds immense promise as a key alternative energy source for achieving decarbonization in the internal combustion engine industry. Despite its advantages, ammonia engines currently face challenges such as low combustion efficiency, combustion instability, and high N_2O emissions. Overcoming these hurdles to achieve stable, efficient, and clean ammonia combustion remains a significant challenge (Liu 2024).

The introduction of ammonia as a fuel in internal combustion engines is envisioned as a promising step towards replacing fossil fuels in ships, heavy-duty vehicles, and stationary power generation engines in the coming years. The driving forces behind this shift are the potential for carbon dioxide emission reductions and ammonia's ability to be produced from renewable energy sources via water electrolysis and the Haber-Bosch process, effectively serving as a chemical energy storage medium.

2.3.2 Conceptual Fuel Injection Mechanisms

Two primary mechanisms exist for ADDF combustion: Low-Pressure Dual Fuel (LPDF) Mode and High-Pressure Dual Fuel (HPDF) Mode. In LPDF Mode, as depicted in Fig. 6a, ammonia is injected into the intake manifold of high-speed four-stroke engines or directly into the cylinders of low-speed marine two-stroke engines with a unidirectional scavenging system during the exhaust stroke at relatively low pressure. The ammonia then mixes with air to form a premixed combustible mixture during the compression stroke. Combustion is subsequently initiated by injecting high-reactivity diesel pilot fuel into the cylinder at an appropriate timing. In this mode, if the diesel pilot fuel share is low, the flame propagation speed of the premixed turbulent flame controls the in-cylinder combustion processes, similar to conventional diesel engines (Zhou 2023). In contrast, HPDF Mode, as illustrated in Fig. 6b, involves the sequential injection of diesel pilot fuel and liquid ammonia into the cylinder near top dead center. In this mode, akin to conventional diesel engines, the rate of fuel-air mixing predominantly influences the in-cylinder heat release processes (Zhou 2023).

A study by Tai Li et al. (2022) numerically investigated and compared the combustion and emission characteristics of Low-Pressure Dual Fuel (LPDF) and High-Pressure Dual Fuel (HPDF) modes. The findings indicate that in LPDF mode, a maximum ammonia energy ratio of around 80% is recommended, as further increases may lead to incomplete combustion. Conversely, HPDF mode has the potential to achieve up to 97% diesel replacement. Compared to pure diesel mode, HPDF mode maintains similar thermal efficiency, cooling loss, and output while significantly reducing equivalent CO2 (i.e., $CO_2 + 300 \times N_2O$) with a slight increase in unburned ammonia emissions. LPDF mode, with reduced heat transfer losses, shows higher potential for achieving greater thermal efficiency than pure diesel mode; however, its capacity to reduce equivalent CO_2 is limited as N_2O emissions increase significantly if after treatment methods are not used (Li et al. 2022). The comparison between LPDF and HPDF modes reveals that the LPDF mode can achieve higher thermal efficiency, while the HPDF mode substantially reduces engine-out emissions of unburned ammonia, NOx, and greenhouse gases. Notably, HPDF mode offers better utilization of inducted air, eliminating the need to premix ammonia with air at the inlet and leading to significantly lower unburned ammonia emissions from trapped gases in crevices. Additionally, HPDF mode provides greater flexibility in the injection strategy, allowing more control over combustion and optimization of engine performance and emissions (Bjørgen et al. 2024).

2.3.3 Research and Development in CI Engines

Fuel composition plays a critical role in shaping the combustion process within internal combustion engines. Ammonia introduces unique challenges due to its

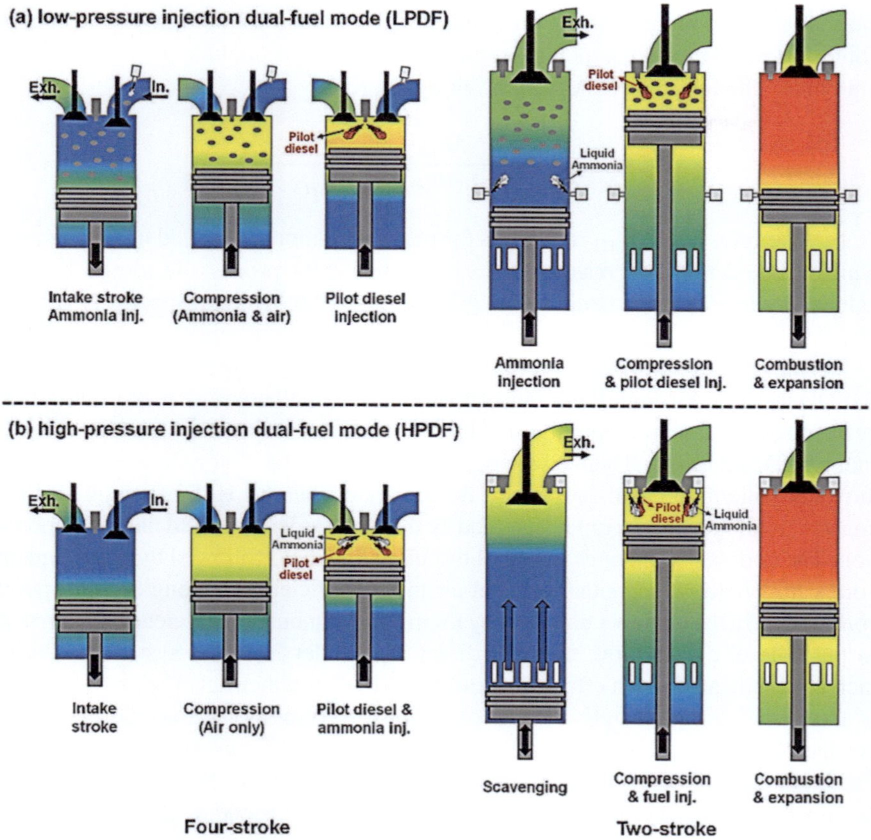

Fig. 2.6 Conceptual models of the pilot diesel-ignited ammonia combustion engine under both the low- and high-pressure injection dual-fuel modes (Zhou 2023)

low flame propagation speed, high autoignition temperature, and the potential for increased NOx and unburned ammonia emissions. To tackle these issues and optimize combustion performance in ADDF engines, researchers have conducted extensive studies to explore the effects of various combustion parameters.

Ammonia Energy Ratio

The ammonia energy ratio is crucial in determining the combustion efficiency, emissions, and overall performance of ammonia/diesel dual-fuel (ADDF) engines. Higher ammonia ratios can lead to significant reductions in carbon emissions but may also introduce challenges such as incomplete combustion and increased NOx formation. Conversely, lower ammonia ratios can result in reduced engine efficiency and diminished environmental benefits. Therefore, optimizing the ammonia-to-diesel ratio is essential for each specific engine type and operating condition. The ammonia energy ratio (AER) is a metric used to quantify the proportion of energy derived from ammonia within the overall fuel mixture. It is calculated as the ratio of the energy content of ammonia to the total energy content of all fuels, including both diesel and ammonia. The formula for AER is given by Eq. (2.1):

$$AER = \frac{M_{NH3} \times 18.8}{M_{diesel} \times LHV_{diesel} + M_{NH3} \times 18.8} \quad (2.1)$$

In this expression, M_{NH3} represents the mass of ammonia involved in combustion, while M_{diesel} denotes the mass of diesel consumed in the process. The lower heating values for diesel and ammonia are 43.4 MJ/kg and 18.8 MJ/kg, respectively (Zheng 2024).

Zheng Zhang et al. (2024) proposed an innovative high-pressure direct injection (HPDI) strategy for liquid ammonia to address the challenges of poor combustion performance and increased unburned ammonia emissions in ammonia/diesel dual-fuel (ADDF) engines. Their investigation into the combustion process in a heavy-duty ammonia/diesel dual-fuel engine demonstrated that this HPDI strategy, under a constant total fuel energy input, can modify the in-cylinder flow field during combustion. The approach generates an initial turbulent flow field followed by a subsequent vortex flow field, which enhances fuel-air mixing efficiency, leading to multi-point combustion in the chamber and thereby improving combustion efficiency. Compared to pure diesel combustion, the dual direct in-cylinder injection system results in increased indicated mean effective pressure (IMEP) and thermal efficiency. The use of ammonia fuel significantly reduces greenhouse gas emissions, and NOx emissions are also lowered at high ammonia energy ratios due to the De-NOx thermal reaction. The main thermal De-NOx reactions involve the conversion of NOx back into harmless nitrogen gas and oxygen. This can be achieved by introducing a reducing agent, such as ammonia, at high temperatures. Changes in combustion characteristics impact both engine performance and pollutant formation, as illustrated in Fig. 2.7, which compares average in-cylinder temperature curves and NOx emissions for different ammonia percentages.

Figure 2.8 illustrates the reduction in greenhouse gas emissions as the ammonia energy ratio (AER) increases. Compared to pure diesel combustion, greenhouse gas emissions and NOx are reduced by 74.3% and 59.9%, respectively, under an 80% ammonia scheme. Although higher AERs significantly reduce greenhouse gas emissions, they also lead to increased unburned ammonia emissions. Therefore, optimizing the AER is essential to balance environmental benefits with combustion efficiency and emission control.

A study by Fatih Okumus (2024) and colleagues explores the effects of ammonia energy ratio (AER) and compression ratio on engine performance and exhaust emissions, offering innovative insights for optimizing engine parameters to enhance performance and reduce emissions. Their findings reveal that engine power decreases with increasing AER, reaching a minimum of 4.53 kW at 60% AER, as shown in Fig. 2.9. In contrast, increasing the compression ratio generally leads to higher engine power. Thermal efficiency improves up to 40% AER but then declines. CO_2 emissions are significantly reduced with higher AER, reaching their lowest levels at 60% AER.

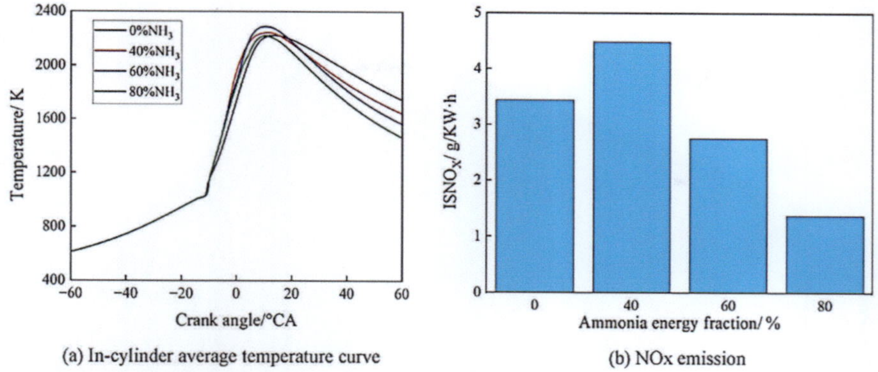

Fig. 2.7 Comparison of average in-cylinder temperature and NOx emissions under different ammonia energy ratio schemes (Zhang 2024)

Fig. 2.8 Comparison of greenhouse gas emissions under different ammonia percentage schemes (Zhang 2024)

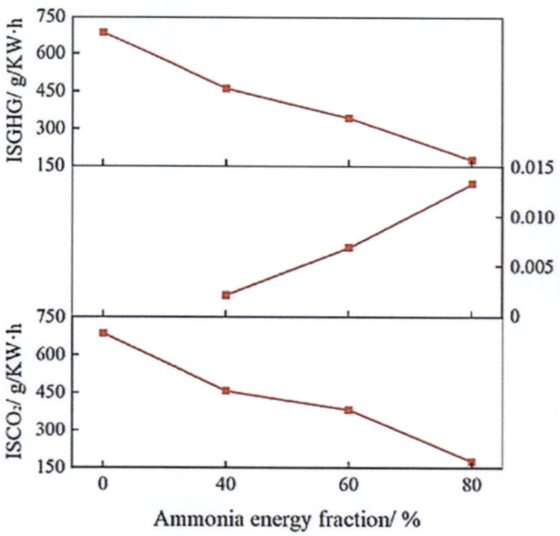

NOx emissions increase with higher ammonia content up to 40%, and increasing the compression ratio further reduces CO_2 emissions while increasing NOx emissions. The study underscores the complex interplay between AER, compression ratio, and engine performance and emissions, highlighting the need to optimize these parameters to balance power output, fuel efficiency, and environmental impact in ammonia-diesel engines.

Injection Methods

This section explores the examination and comparison of various ammonia injection methods, including direct and staged injection, and their impact on engine performance and emissions. In dual-fuel ammonia-diesel engines, the diesel fuel injection

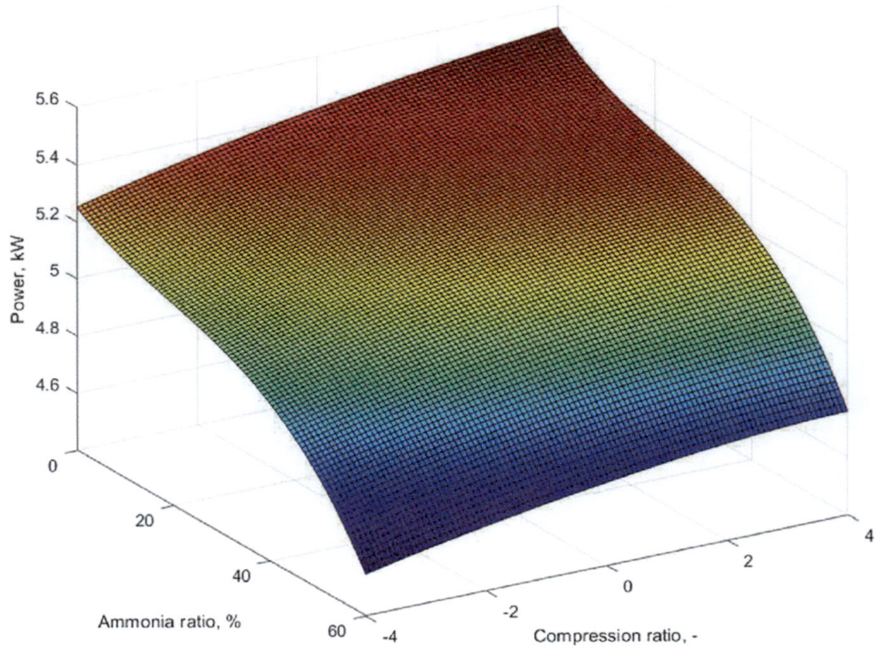

Fig. 2.9 Effects of AER and compression ratio on engine power (Okumuş et al. 2024)

strategy is crucial for optimizing combustion and emissions. An effective injection strategy not only improves engine emission performance but also ensures efficient fuel combustion within the combustion chamber. Therefore, investigating ammonia/diesel fuel injection strategies is of significant importance. Existing research has focused on modifying the combustion process using techniques such as early injection or two-stage injection. Despite these efforts, achieving lower unburned ammonia emissions while maintaining optimal engine performance continues to be a challenge (Shi 2024). Research has shown that ammonia/diesel premixed combustion offers improved efficiency and reduced greenhouse gas emissions compared to conventional diesel combustion. However, it also results in significant levels of NOx and unburned ammonia. Therefore, the direct injection strategy for liquid ammonia requires further investigation. This calls for comprehensive analyses of the combustion process and pollutant formation to optimize performance and minimize emissions (Shin and Park 2024).

In a study by Shin (2024), a numerical analysis of the ammonia/diesel direct injection (ADDI) combustion process was conducted to investigate an optimal injection strategy. The study found that compensating for ammonia's slow flame speed by providing additional momentum through direct injection can enhance combustion characteristics. Direct injection allows ammonia to reach the high-temperature zone

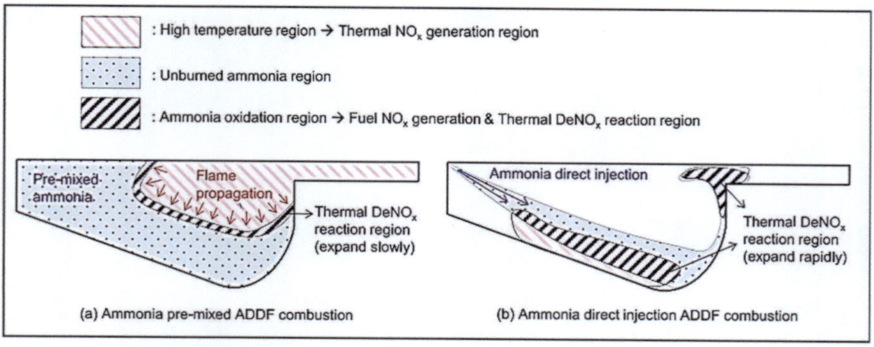

Fig. 2.10 Schematics of thermal De-NO_x zone formation in ADDF combustion **a** premixed and **b** direct injection (Shin and Park 2024)

created by diesel combustion, overcoming its inherent limitation of low flame propagation speed. Additionally, the expanded thermal De-NOx reaction zone in ADDI combustion compared to premixed ammonia/diesel dual-fuel (ADDF) combustion facilitates more effective NO removal. Continuous introduction of NH_2, an ammonia oxidation product, into the NO production zone enables NO removal across a broader region than in ADDF premixed combustion, which has a narrower thermal De-NOx reaction zone ahead of the flame, as shown in Fig. 2.10. Consequently, ADDI combustion is more efficient in reducing NOx emissions and significantly reduces unburned ammonia compared to ADDF premixed combustion.

If the ammonia injection timing is earlier than the diesel injection timing, the combustion efficiency will be reduced due to the very lean mixture caused by the increased ammonia mixing time or the low temperature area created by the evaporation of ammonia. As can be seen in Fig. 2.11, with the delay of ammonia injection timing, the emission of unburned ammonia increases. However, the maximum combustion temperature and thermal NOx under these conditions are reduced, so the NO emission is reduced. By injecting ammonia at 7° ATDC (after top dead center) and diesel in the range of 15° to 10° ATDC, the efficiency is increased by 8%, NO emissions are reduced by up to 13.5%, and greenhouse gas emissions are reduced by about 91% compared to pure diesel mode.

In a study by Liu et al. (2024), the impact of ammonia energy ratio (AER) and diesel split ratio (DSR) on combustion and emissions in an ammonia/diesel dual-fuel optical engine with two-stage injection was examined. The findings reveal that increasing the AER to 60% significantly reduces peak cylinder pressure, peak apparent heat release rate (AHRR), and the rate of cylinder pressure rise, along with a decrease in indicated mean effective pressure (IMEP) from 4.16 bar to 3.24 bar, indicating lower combustion efficiency with higher ammonia content. This reduction is attributed to decreased soot formation and limitations in flame area and propagation speed, largely due to ammonia's slower combustion rate and its chemical interaction with diesel. The two-stage injection system, particularly the role of the second

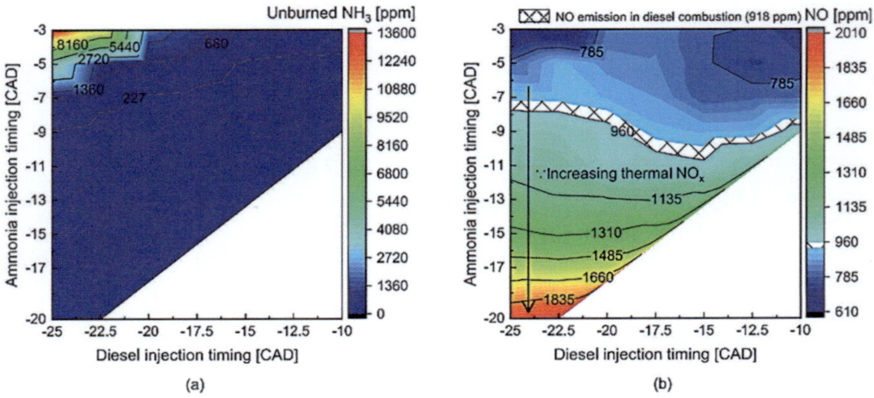

Fig. 2.11 The emissions of **a** unburned ammonia and **b** NO for different ammonia and diesel injection timings at an ammonia energy ratio of 97% (Shin and Park 2024)

injection, enhances peak cylinder pressure and AHRR. As depicted in Fig. 2.12, optimizing the DSR to 40% improves combustion efficiency by enhancing fuel-oxygen interaction. However, DSR values of 60% and 80% pose challenges, as the reduced diesel content in the second injection results in a leaner mixture, which adversely affects combustion efficiency, leading to decreased heat release rate (HRR) and a drop in IMEP.

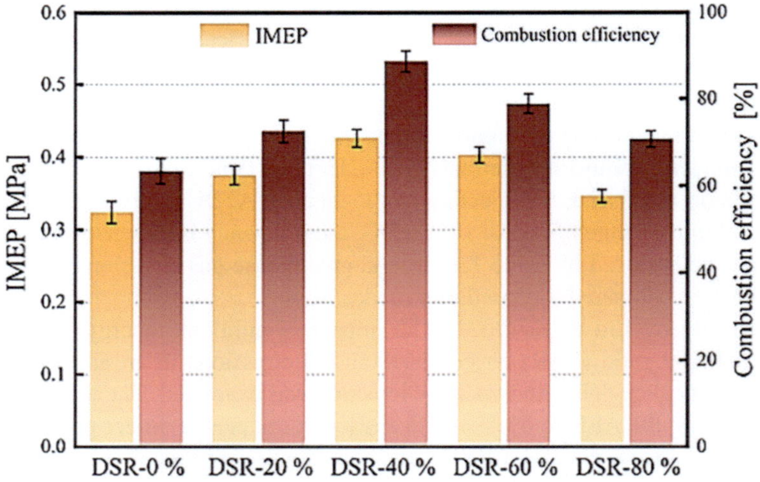

Fig. 2.12 IMEP and combustion efficiency under different primary injection split ratios (DSR) (Liu 2024)

Injection Timing

The timing of ammonia injection into the combustion chamber profoundly affects combustion efficiency, pollutant emissions, and overall engine performance. Early injection can lead to incomplete combustion, while late injection may result in increased soot formation.

Research findings by Wang et al. (2024) indicates that as the ammonia energy ratio (AER) increases, the range of diesel injection required for stable engine operation narrows, and the indicated thermal efficiency (ITE) decreases. Despite this, the ammonia/diesel dual-fuel (ADDF) engine can maintain an indicated mean effective pressure (IMEP) coefficient of variation below 3% across a broad range of AERs. Notably, with optimized diesel injection timing, a theoretical thermal efficiency of approximately 34% is achievable at the highest AER tested (90%). Constant AER optimization of diesel injection timing also reduces unburned ammonia emissions to some extent, though it leads to a gradual increase in NOx emissions. While CO_2 emissions remain relatively stable with changes in diesel injection timing, N_2O emissions vary significantly. Additionally, research by Tian et al. (2024) explores the diffusion combustion of ammonia/diesel blends, investigating parameters such as ammonia injection pressure, the timing of ammonia and diesel injections, diesel injection pressure, and ammonia nozzle diameter. The study found that injecting ammonia onto a diesel flame promotes more complete combustion of ammonia, with pre-injecting ammonia before diesel achieving the highest heat release rate. However, rapid ammonia evaporation can hinder ignition and cause incomplete combustion. Diesel injection pressure does not significantly affect ammonia combustion, but increasing ammonia injection pressure improves fuel-air mixing, thus enhancing the peak heat release rate. Larger ammonia nozzle diameters are necessary for achieving a high AER, though injecting excessive ammonia at once may result in incomplete combustion.

Injection timing is a crucial factor influencing the combustion process, as it directly affects the distribution of diesel fuel within the combustion chamber. This goal is more readily achievable by modifying the injection angle of the diesel nozzle, which significantly impacts both the distribution of fuel and the combustion process itself. However, adjusting the injection angle becomes challenging once the injector is installed. If the engine's fuel composition changes, the original injection angle may become unsuitable. Shi et al. (2024), developed an innovative spray strategy to address this issue by optimizing the timing and angle of diesel injection to ensure a uniform mixture of diesel and ammonia in dual ammonia/diesel engines. Figure 2.13 illustrates the diesel injection angle, denoted by α, which is the angle between the injection direction and the negative Z-axis.

Computational fluid dynamics (CFD) modeling was utilized to assess the performance of a Caterpillar 3401 diesel engine operating on a blend of ammonia and diesel. The findings reveal that a blend with 40% ammonia energy content achieves the lowest emissions while maintaining engine performance. This blending ratio results

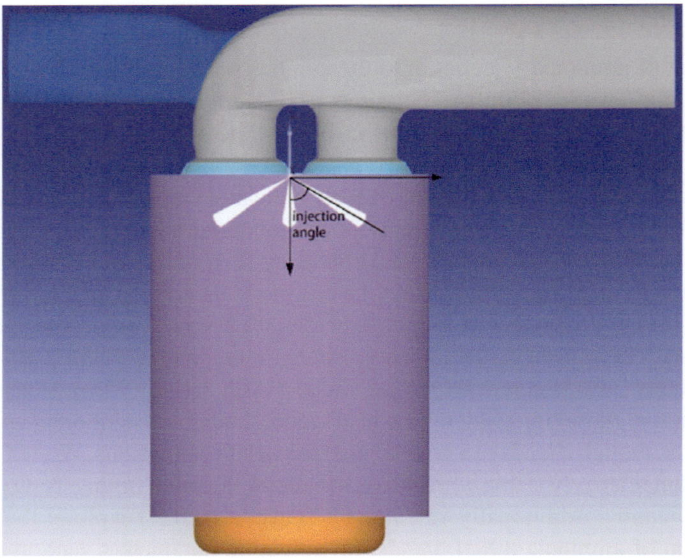

Fig. 2.13 Schematic diagram of diesel injection angle definition (Shi 2024)

in a 34.3% reduction in greenhouse gas emissions compared to pure diesel combustion, with only a 4% decrease in indicated mean effective pressure (IMEP). By optimizing both injection timing and angle, multi-point pilot combustion was achieved, reducing combustion duration and mimicking the effects of premixed combustion. This optimization led to a significant reduction in unburned ammonia emissions, with an emission rate of 0.279 g/kWh—just 1.63% of that from the initial 40% ammonia energy ratio injection scheme. Furthermore, the accelerated combustion enabled by the new injection strategy resulted in higher in-cylinder temperatures, significantly decreasing N_2O and CO emissions (Shi 2024). Table 2.3 provides a comparative analysis of unburned ammonia, greenhouse gas, NOx, and CO emissions across four injection strategies, highlighting the lowest emissions for each pollutant. Detailed emission data for these strategies are illustrated in Fig. 2.14.

Table 2.3 Optimized injection schemes for the four pollutants

Scheme	Parameters	The lowest emissions of pollutants
S1	−62° ATDC injection timing with an 80° injection angle	Unburned NH_3
S2	−30° ATDC injection timing with a 50° injection angle	GHG
S3	−24° ATDC injection timing with an 85° injection angle	NO
S4	−62° ATDC injection timing with an 85° injection angle	CO

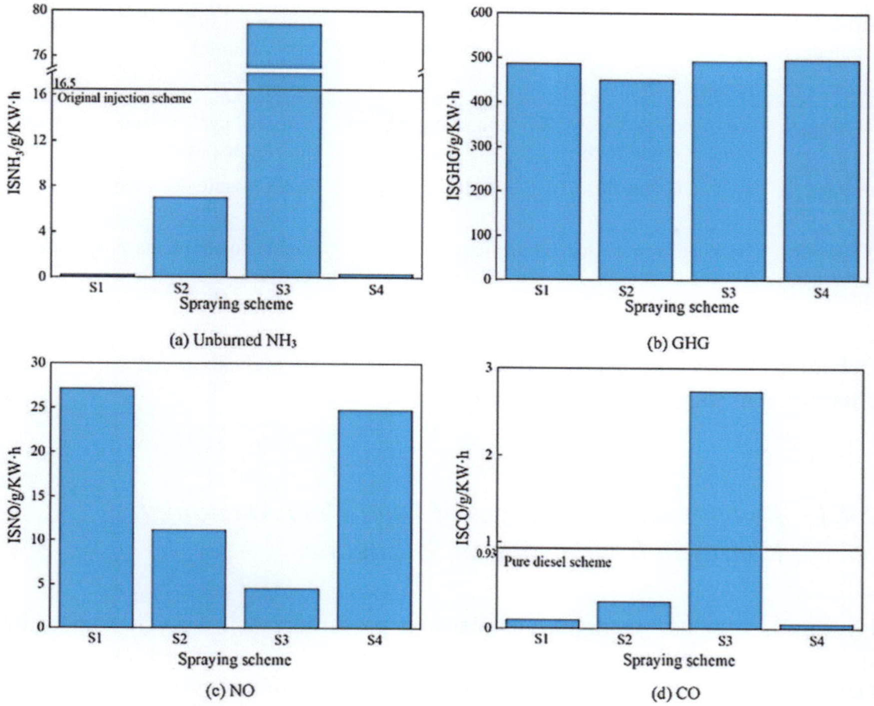

Fig. 2.14 Comparison of emissions for the four schemes with the lowest pollutant emissions (Shi 2024)

2.4 Dual-Fuel Ammonia Combustion in SI Engines

Ammonia's inadequate combustion properties, such as high ignition energy, low heating value, and restricted flammability, require its blending with other fuels. These supplementary fuels function as combustion enhancers, increasing the combustion rate of the ammonia-air flame and boosting the performance of ammonia-powered engines (Newhall and Starkman 1967). In spark-ignition engines, ammonia can be mixed with hydrogen, gasoline, natural gas, and other fuels. As indicated in Table 2.4, incorporating ammonia with other fuels can enhance combustion performance to some extent. However, while this blending reduces emissions of carbon-based pollutants (HC, CO, and CO2), it can also result in increased NOx emissions.

Hydrogen, when blended with ammonia, offers distinct advantages over other fuel options. It significantly enhances combustion performance while minimizing negative effects compared to alternative fuels. Both hydrogen and ammonia have high octane ratings, which enable engines to operate at higher compression ratios, leading to improved thermal efficiency and increased power output. Ideally, this combination aims for zero emissions of additional carbon-based pollutants. Although blended fuels do produce NOx emissions similar to those from fossil fuels, existing

Table 2.4 Comparison of the combustion characteristics of different fuels after mixing with ammonia (compared with the engines using pure fuel in the "Type" column)

Type	Emission effects	Combustion effects
Hydrogen	No carbon emissions, NOx emissions are the same as fossil fuels	Increased cycle thermal efficiency and engine power
Natural gas	NOx, NH_3 increase, CO_2 decrease	Peak pressure decrease
Gasoline	NOx, NH_3 increase, CO decrease	Engine power increase, Pressure decrease

technologies like Selective Catalytic Reduction (SCR) and advanced catalysts are effective in managing these emission (Xu et al. 2022).

2.4.1 *Benefits and Challenges of Dual-Fuel Ammonia/Hydrogen Combustion in SI Engines*

Hydrogen, with its exceptional combustion properties such as high combustion speed, excellent diffusivity, and low ignition energy, is considered an ideal fuel. Ongoing research focuses on producing hydrogen with minimal or zero carbon emissions throughout its entire lifecycle, from production to use. When combined with ammonia, hydrogen improves the combustion process (Xu et al. 2022). Additionally, burning an ammonia/hydrogen blend results in zero carbon-based pollutants, addressing significant environmental concerns and meeting future fuel standards. However, the high temperatures and the need for air mixing during combustion still lead to some nitrogen oxide (NOx) emissions (Xu et al. 2022).

The minimum ignition energy (MIE) is traditionally defined as the smallest amount of energy needed to start combustion (Cui et al. 2016). A lower MIE helps establish a stable ignition core early in the combustion process. Although ammonia has a high hydrogen content, it has a notably high MIE on its own. Even with hydrogen blends and adjusted ignition systems, ammonia/hydrogen mixtures require ignition energy significantly higher by two orders of magnitude compared to hydrocarbon fuels and four orders compared to pure hydrogen (Wang et al. 2021). Experimental studies by Xin et al. (2022) into the combustion and emission characteristics of hydrogen/ammonia fueled engines under partial load conditions showed that ammonia's high MIE increased ignition delay and extended flame development. However, adjusting the ignition timing allowed for improvements in thermal efficiency while keeping NOx emissions within acceptable limits (Fig. 2.15).

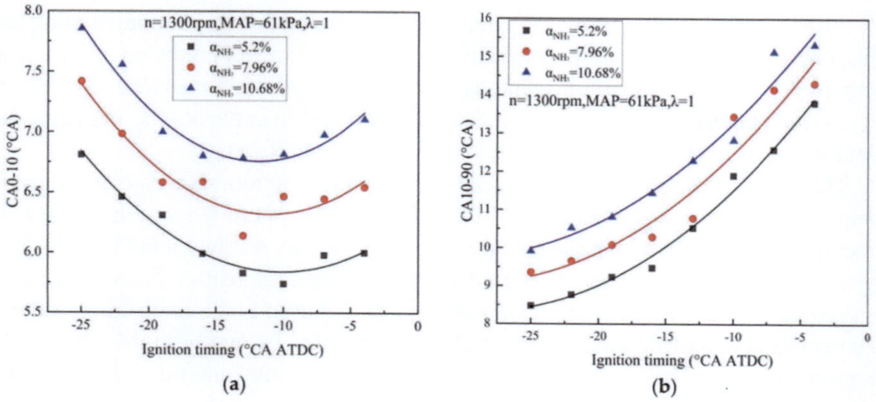

Fig. 2.15 Combustion durations relative to ignition timing for different ammonia concentrations. **a** CA0-10 duration for various ammonia additions **b** CA10-90 duration for various ammonia additions (Xin et al. 2022)

2.4.2 Ammonia as Hydrogen Carrier

Ammonia/hydrogen engines necessitate complex fuel delivery systems because hydrogen and ammonia must be stored and managed separately (Xu et al. 2022). As shown in Fig. 2.16, fuels mixed with ammonia offer better combustion thermal efficiency compared to pure gasoline. The efficiency of combustion is affected by the blend ratio, with ammonia/hydrogen mixtures generally achieving the best performance when the hydrogen volume fraction is around 10% (Lesmana et al. 2019).

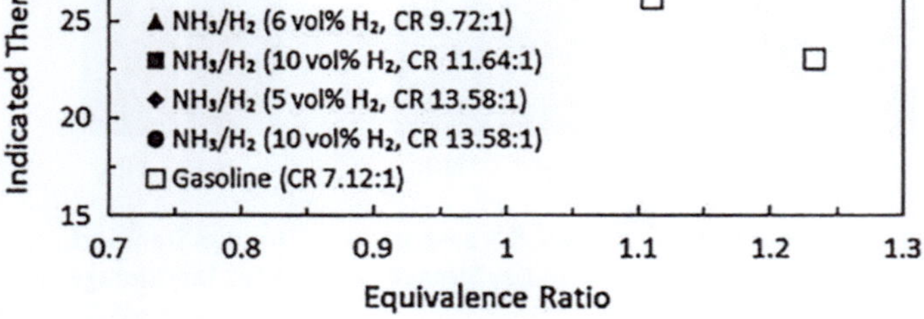

Fig. 2.16 Thermal efficiency across various compression ratios and hydrogen doping levels (Lesmana et al. 2019)

Although the amount of hydrogen needed for combustion in ammonia/hydrogen engines is relatively modest, generating hydrogen from ammonia involves extra expenses. However, using ammonia as a carrier for hydrogen provides significant economic benefits for storage and transport compared to conventional methods, as demonstrated in Fig. 2.17 (Bartels 2008; Ahamed et al. 2023).

The decomposition of ammonia is an endothermic reaction, requiring an external heat source to proceed. Ryu et al. (2014) designed a system where ammonia was decomposed directly within the exhaust pipe of an engine. As shown in Fig. 2.18, the heat generated by the exhaust was used to power the decomposition process. A catalyst containing 2% ruthenium on alumina pellets was used to facilitate this reaction. The catalyst was placed in a cylindrical tube, preceded by another tube filled with heat-exchanging wiring. After decomposition, the resulting mixture of hydrogen,

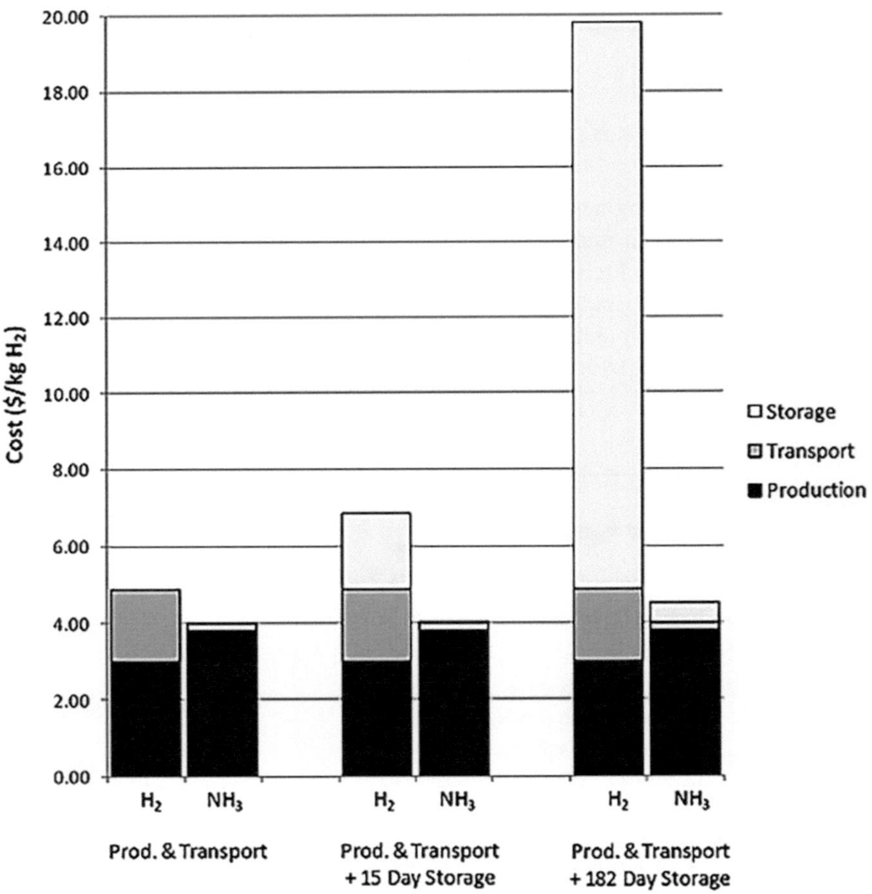

Fig. 2.17 Cost comparison of hydrogen and ammonia production, transportation, and storage (Bartels 2008) storage

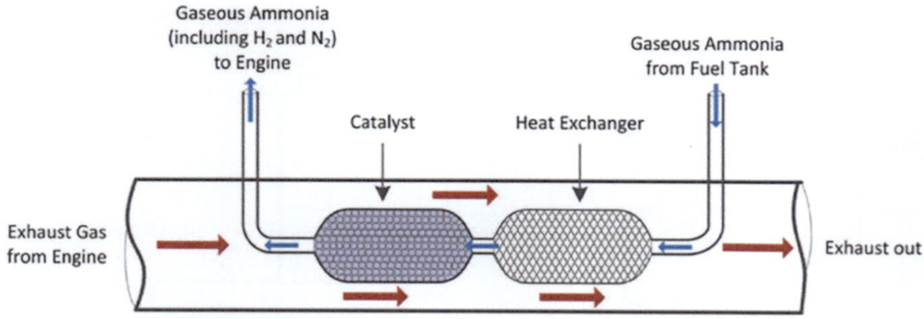

Fig. 2.18 Schematic of ammonia decomposer assembly (Ryu et al. 2014)

nitrogen, and any remaining ammonia was injected into the engine cylinder and mixed with gasoline. The study found that the efficiency of ammonia decomposition was influenced by the rate at which ammonia was introduced to the catalyst.

Comotti and Frigo (2015) developed a more intricate ammonia decomposition system that harnesses engine exhaust heat, as shown in Fig. 2.19. Using a commercial Ru-based catalyst, they achieved a 98% ammonia conversion rate and a hydrogen production of 1.4 Nm^3/h. The produced hydrogen was then mixed with ammonia and injected into the engine. This ammonia/hydrogen combustion strategy resulted in thermal efficiency comparable to gasoline, with low combustion variability. The researchers demonstrated the feasibility of operating a spark ignition engine on a mixture of ammonia and on-board produced hydrogen, using ammonia as the primary fuel source. A built-in electric heater was included to facilitate cold starts.

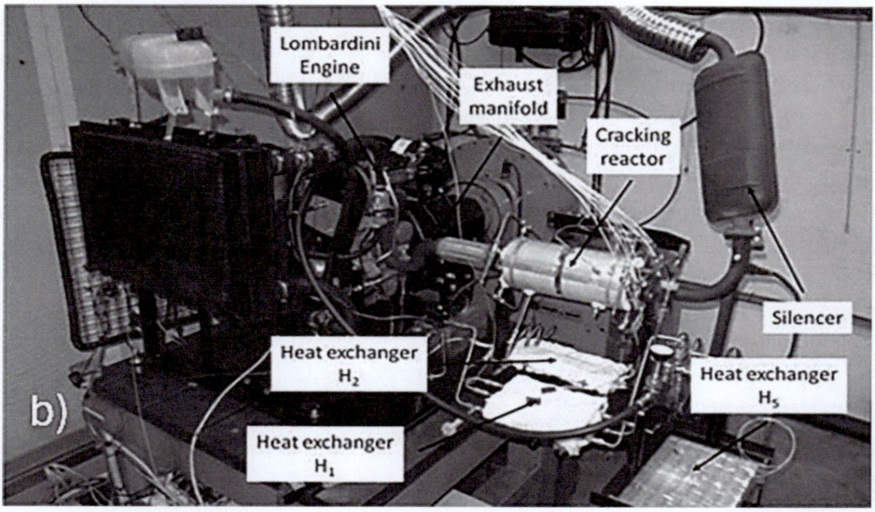

Fig. 2.19 A hydrogen generation system from ammonia decomposition using engine exhaust heat (Comotti and Frigo 2015)

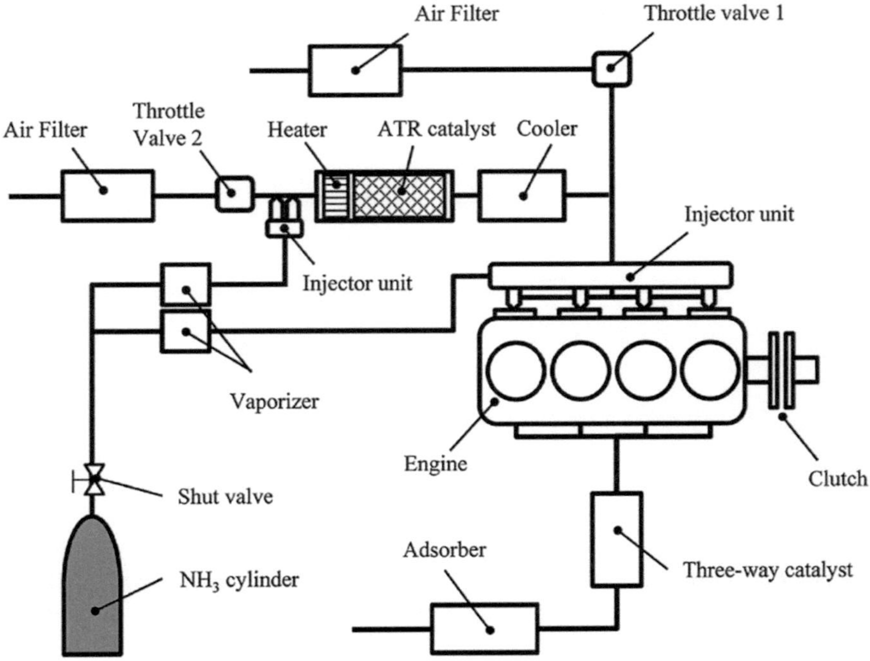

Fig. 2.20 Schematic of hydrogen production from ammonia reforming (Koike et al. 2021)

The aforementioned studies were conducted in an oxygen-free environment. To compensate for the endothermic nature of ammonia decomposition, longer electrical heating might be necessary during cold starts to achieve optimal hydrogen production temperatures. Koike et al. (2021) investigated an alternative ammonia reforming approach, incorporating a catalyst and electric heating unit, as shown in Fig. 2.20. The primary function of the electric heater was to preheat the catalyst during cold starts. Unlike previous methods, Koike et al. (2021) introduced a small amount of air into the reforming process. This oxygen presence resulted in a partial combustion of ammonia, releasing heat to sustain the catalyst temperature for efficient ammonia decomposition. The reformed gas was then injected into the engine intake for mixing with ammonia. With an equivalent ratio of 5.4 in the reformer, engine startup was achieved within 3 seconds. The hydrogen fraction was initially 15% at the reformer outlet and decreased to less than 10% in the cylinder. After 8 seconds, the hydrogen fraction stabilized at nearly 50% at the reformer outlet.

2.4.3 Research and Development in SI Engines

Initial efforts to use ammonia as a fuel for spark-ignition engines began in the late 1960s for military applications, aiming primarily to achieve fuel independence from hydrocarbon sources (Cornelius et al. 1966). Garabedian and Johnson (Tornatore et al. 2022) conducted experimental research to assess the performance of a 48 kW, 4-cylinder spark-ignition engine using anhydrous ammonia as fuel. Initial tests with an LPG carburetor to deliver ammonia resulted in a substantial decrease in maximum rated power to 7 kW and engine speed to 2000 rpm. Introducing small amounts of hydrogen (2–4%) increased power output to around 20 kW and extended the speed range to 3200 rpm. Further improvements were made by raising the compression ratio, leveraging ammonia's high-octane number, ultimately achieving a power output of 40 kW at 4000 rpm.

Multiple studies have been performed to optimize the hydrogen content in ammonia/hydrogen fuel blends for spark-ignition engines. Mørch et al. (2011) examined the performance of a CFR engine using various ammonia/hydrogen mixtures, air-fuel ratios, and compression ratios. Their results showed that a hydrogen volume fraction of about 20% maximized engine efficiency at 1000 rpm with a compression ratio of 10 (Fig. 2.21). Frigo and Gentili (2013) proposed that engine load was the primary factor in determining the optimal hydrogen content in ammonia/hydrogen fuel blends, whereas engine speed was less influential. Lhuillier et al. (2019) explored the effects of adding hydrogen to an ammonia-fueled SI engine. They varied the hydrogen content by volume from 5% to 15% and tested different air-fuel mixtures and intake pressures. Adding hydrogen enhanced combustion stability across various engine operating conditions. The optimal hydrogen content was found to be 10%, which improved combustion efficiency without adversely affecting emissions. The highest engine efficiency was achieved with slightly lean fuel mixtures. However, further leaning the mixture increased NOx and ammonia emissions without enhancing efficiency. These findings are consistent with observations by Frigo and Gentili (2013).

Lhuillier et al. (2020) conducted an extensive study on the performance of a single-cylinder spark-ignition engine using ammonia/hydrogen blends with various equivalence ratios and hydrogen volume fractions up to 60%. As shown in Fig. 2.22, indicated efficiency was found to be influenced by both the equivalence ratio and hydrogen content. Consistent with previous studies, optimal indicated efficiency was achieved with moderate hydrogen concentrations between 5% and 20% under slightly lean conditions. The highest efficiency, around 39%, was obtained with a 20% hydrogen volume fraction at an equivalence ratio of 0.9. Excessively high hydrogen concentrations led to reduced efficiency due to increased heat transfer losses from higher flame temperatures. Dinesh et al. (2022) experimentally investigated the impact of hydrogen content (5–21 vol%) on engine performance across various compression ratios (12–15) and engine speeds (1400 and 1700 rpm). Their results showed that a hydrogen fraction of 21%, combined with a compression ratio of 15 and an engine speed of 1400 rpm, produced the best overall engine performance. Pyrc et al. (2023) conducted experiments demonstrating that a hydrogen energy fraction of 12% within the fuel mixture effectively mitigated ignition instability at compression ratios of 8 and 10. Their studies explored a hydrogen energy fraction range of 0–70%.

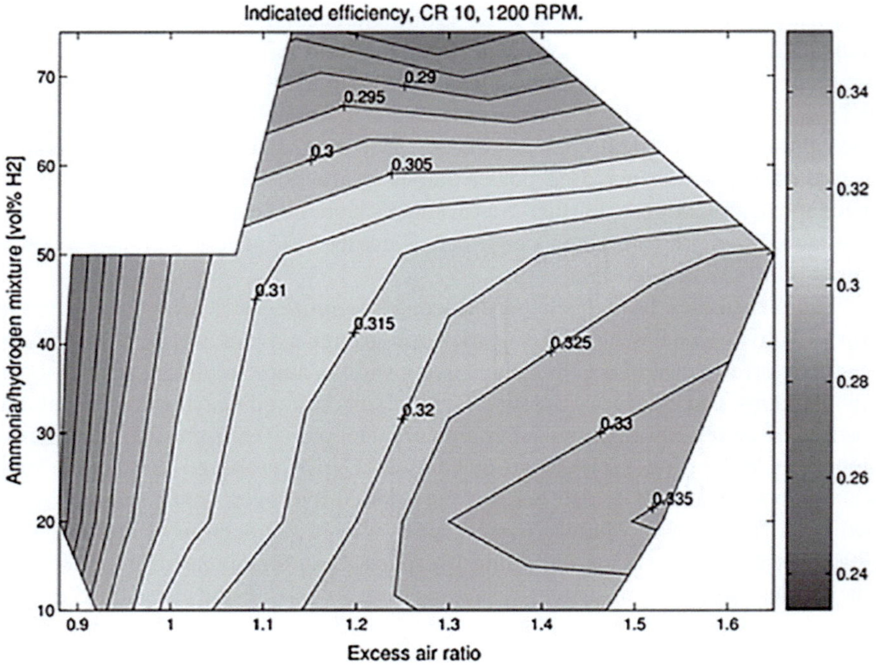

Fig. 2.21 Thermal efficiency of a spark-ignition ammonia/hydrogen engine as it changes with equivalence ratio and hydrogen content (Mørch et al. 2011) content

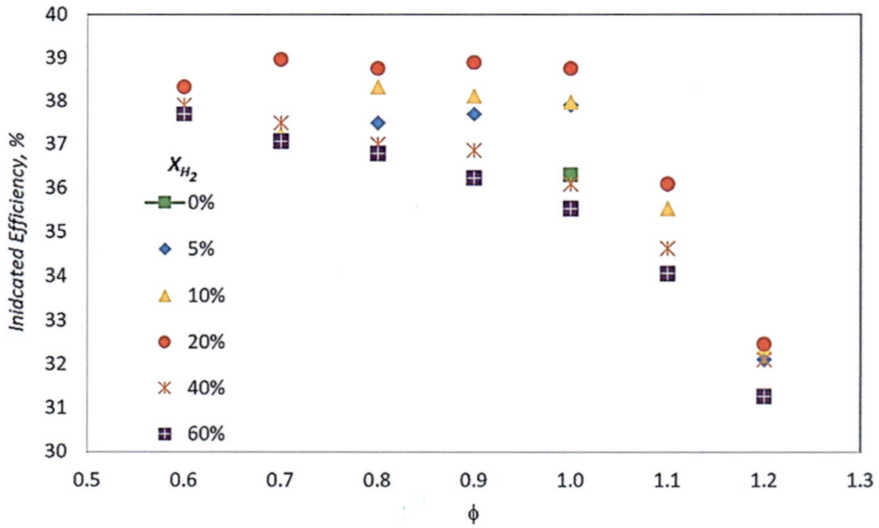

Fig. 2.22 Indicated efficiency in relation to equivalence ratio (ϕ) at varying hydrogen concentrations (Lhuillier et al. 2020)

The performance of ammonia/hydrogen engines depends greatly on operational parameters such as air-fuel mixture ratio and engine compression ratio. Wang et al. (2021) conducted a numerical investigation into the combustion characteristics and emissions of ammonia/hydrogen blends across different conditions. Their study revealed that combustion efficiency decreased with leaner air-fuel mixtures due to incomplete fuel oxidation, particularly noticeable with higher hydrogen content. However, ammonia/hydrogen blends exhibited excellent resistance to knocking, which enhances their suitability for higher compression ratios.

To explore the impact of compression ratio on engine performance, Dinesh and Kumar (2022) carried out experimental tests with different compression ratios and engine speeds, keeping ignition timing constant. Their results showed a positive correlation between compression ratio and combustion rate, evidenced by shorter combustion duration, higher peak cylinder pressure, increased temperature, and greater heat release rate. Fredrik R. Westlye et al. (2013) investigated NOx emissions from an ammonia/hydrogen engine with a 0.612-l displacement across different compression ratios. They discovered that higher compression ratios led to increased levels of unburned ammonia. Ignition timing influenced nitrous oxide (N_2O) production, with delayed ignition resulting in higher N_2O emissions. Optimal ignition timing helped minimize N_2O emissions. Although the mechanisms differ, NOx emissions from ammonia/hydrogen combustion are similar to those from conventional fossil fuels, making technologies like selective catalytic reduction (SCR) effective in reducing NOx emissions from ammonia/hydrogen engines.

Ji et al. (2024) conducted a numerical study to examine the impact of high compression ratios (CRs) on ammonia/hydrogen engine performance. The study included four CRs from 10.7 to 13.7 and varied spark timing (ST) between 28°CA and 0°CA (BTDC). Higher CRs improved ignition stability, reduced combustion duration, decreased cooling losses, and increased power output, but also led to higher NOx emissions. Optimal performance at high CRs was achieved by carefully adjusting ST, resulting in peak IMEP and ITE values of 4 bar and 38.3%, respectively. The optimal spark timing for maximum braking torque (MBT) shifted closer to top dead center as CR increased. Additionally, Lhuillier et al. (2019) highlighted the potential of supercharging to improve the indicated efficiency of ammonia/hydrogen engines.

2.5 Conclusion

Ammonia, a sustainable and carbon-free fuel, shows great promise for internal combustion engines. While facing combustion challenges, ongoing research is focusing on optimizing its use through strategies like dual-fuel combustion and blending with other fuels. Ammonia's advantages, including high energy density and existing infrastructure, make it a potential solution for decarbonizing transportation. With continued development, ammonia could play a significant role in transitioning to a cleaner energy future. In the maritime industry, ammonia emerges as a promising alternative fuel due to its potential for reduced greenhouse gas emissions and high energy density. While significant technical and economic hurdles must be overcome, ongoing research and development efforts are essential to unlock the full potential of ammonia as a sustainable marine fuel. As the industry continues to explore and invest in innovative solutions, a cleaner and more sustainable future for maritime transportation is increasingly within reach.

Ammonia dual-fuel combustion offers several benefits, including reduced fossil fuel reliance, easier implementation, and improved engine performance. In CI engines, by substituting a portion of diesel with ammonia, pollutant emissions and crude oil dependency can be decreased. Existing diesel engines can be modified to accommodate ammonia, and ammonia can enhance combustion efficiency and reduce fuel consumption. While NOx emissions remain a concern, existing technologies can effectively manage them. In SI engines, hydrogen is a particularly promising blending partner for ammonia. It offers significant benefits in terms of combustion performance and emission reduction. Both hydrogen and ammonia have high octane ratings, enabling engines to operate at higher compression ratios and improve efficiency. Also, ammonia can be used as a hydrogen carrier, offering benefits in storage and transportation. While ammonia itself requires blending with other fuels for efficient combustion, the hydrogen produced from ammonia can be used to enhance combustion performance.

While ammonia holds promise as a sustainable alternative fuel, challenges such as low combustion efficiency, instability, and high N_2O emissions need to be addressed for widespread adoption. Overcoming these hurdles is crucial for achieving a cleaner and more sustainable future for internal combustion engines. Overall, ammonia presents a promising alternative fuel for internal combustion engines, offering a potential solution to reduce greenhouse gas emissions and dependence on fossil fuels. With continued research and development, ammonia could play a significant role in decarbonizing transportation and contributing to a cleaner and more sustainable future.

References

Ahamed M et al (2023) Comparative assessment of sCO2 cycles, optimal ORC, and thermoelectric generators for exhaust waste heat recovery applications from heavy-duty diesel engines. Energies 16(11):4339

Ahbabi Saray J, Gharehghani A, Hosseinzadeh D (2024) Towards sustainable energy carriers: a solar and wind-based systems for green liquid hydrogen and ammonia production. Energy Convers Manag 304:118215

Bartels JR (2008) A feasibility study of implementing an ammonia economy. Iowa State University

Bjørgen KOP, Emberson DR, Løvås T (2024) Combustion of liquid ammonia and diesel in a compression ignition engine operated in high-pressure dual fuel mode. Fuel 360

Cardoso JS et al (2021) Ammonia as an energy vector: current and future prospects for low-carbon fuel applications in internal combustion engines. J Clean Prod 296:126562

Cha J et al (2021) An efficient process for sustainable and scalable hydrogen production from green ammonia. Renew Sustain Energy Rev 152:111562

Comotti M, Frigo S (2015) Hydrogen generation system for ammonia–hydrogen fuelled internal combustion engines. Int J Hydrogen Energy 40(33):10673–10686
Cornelius W, Huellmantel LW, Mitchell HR (1966) Ammonia as an engine fuel. SAE Trans 74:300–326
Cui G et al (2016) Experimental study of minimum ignition energy of methane/air mixtures at elevated temperatures and pressures. Fuel 175:257–263
Cunanan C et al (2021) A review of heavy-duty vehicle powertrain technologies: diesel engine vehicles, battery electric vehicles, and hydrogen fuel cell electric vehicles. Clean Technol 3(2):474–489
Di Blasio G et al (2022) Clean fuels for mobility. Springer
Di Micco S et al (2024) Ammonia-powered ships: concept design and feasibility assessment of powertrain systems for a sustainable approach in maritime industry. Energy Convers Manag: X 22:100539
Dimitriou P, Javaid R (2020) A review of ammonia as a compression ignition engine fuel. Int J Hydrogen Energy 45(11):7098–7118
Dinesh M, Kumar G (2022) Effects of compression and mixing ratio on NH3/H2 fueled Si engine performance, combustion stability, and emission. Energy Convers Manag: X 15:100269
Dinesh M, Pandey JK, Kumar G (2022) Study of performance, combustion, and NOx emission behavior of an SI engine fuelled with ammonia/hydrogen blends at various compression ratio. Int J Hydrogen Energy 47(60):25391–25403
Frigo S, Gentili R (2013) Analysis of the behaviour of a 4-stroke Si engine fuelled with ammonia and hydrogen. Int J Hydrogen Energy 38(3):1607–1615
Ghandehariun S, Gharehghani A, Saray JA (2024) Chapter eighteen—ethylene and ethanol production from CO_2. In: Rahimpour MR, Makarem MA, Meshksar M (eds) Advances and technology development in greenhouse gases: emission, capture and conversion. Elsevier, pp 427–442
Giddey S, Badwal S, Kulkarni A (2013) Review of electrochemical ammonia production technologies and materials. Int J Hydrogen Energy 38(34):14576–14594
Gray Jr JT et al (1967) Ammonia fuel—engine compatibility and combustion. SAE Trans 785–807
Ji C et al (2024) Numerical investigation on the combustion performance of ammonia-hydrogen spark-ignition engine under various high compression ratios and different spark-ignition timings. Int J Hydrogen Energy 56:817–827
Koike M et al (2021) Cold-start performance of an ammonia-fueled spark ignition engine with an on-board fuel reformer. Int J Hydrogen Energy 46(50):25689–25698
Kumar S et al (2024) Ammonia and hydrogen for green energy transition. Springer
Lee J et al (2010) Studies on properties of laminar premixed hydrogen-added ammonia/air flames for hydrogen production. Int J Hydrogen Energy 35(3):1054–1064
Lee J et al (2024) Strategic pathways to alternative marine fuels: empirical evidence from shipping practices in South Korea. Sustainability 16:2412
Lesmana H et al (2019) NH3 as a transport fuel in internal combustion engines: a technical review. J Energy Res Technol 141(7):070703
Lhuillier C et al (2020) Experimental study on ammonia/hydrogen/air combustion in spark ignition engine conditions. Fuel 269:117448
Lhuillier C et al (2019) Performance and emissions of an ammonia-fueled SI engine with hydrogen enrichment. SAE Technical Paper
Li T et al (2022) A comparison between low- and high-pressure injection dual-fuel modes of diesel-pilot-ignition ammonia combustion engines. J Energy Inst 102:362–373
Liu X et al (2024) Optical diagnostics in impact of ammonia energy ratio and diesel split ratio on combustion process and flame propagation in an Ammonia-diesel Dual-Fuel engine. Fuel 364
Mørch CS et al (2011) Ammonia/hydrogen mixtures in an SI-engine: engine performance and analysis of a proposed fuel system. Fuel 90(2):854–864
Musculus MP, Miles PC, Pickett LM (2013) Conceptual models for partially premixed low-temperature diesel combustion. Prog Energy Combust Sci 39(2–3):246–283

Newhall HK, Starkman ES (1967) Theoretical performance of ammonia as a gas turbine fuel. SAE Trans 75:772–784

Okumuş F et al (2024) The effects of ammonia addition on the emission and performance characteristics of a diesel engine with variable compression ratio and injection timing. Int J Hydrogen Energy 64:186–195

Pearsall TJ, Garabedian CG (1968) Combustion of anhydrous ammonia in diesel engines. SAE Trans 3213–3221

Pyrc M et al (2023) Assessment of the co-combustion process of ammonia with hydrogen in a research VCR piston engine. Int J Hydrogen Energy 48(7):2821–2834

Ryu K, Zacharakis-Jutz GE, Kong S-C (2014) Performance enhancement of ammonia-fueled engine by using dissociation catalyst for hydrogen generation. Int J Hydrogen Energy 39(5):2390–2398

Sagel VN, Rouwenhorst KH, Faria JA (2022) Green ammonia enables sustainable energy production in small island developing states: a case study on the island of Curaçao. Renew Sustain Energy Rev 161:112381

Senecal P, Leach F (2019) Diversity in transportation: why a mix of propulsion technologies is the way forward for the future fleet. Results Eng 4:100060

Shi C et al (2024) Parametric analysis and optimization of the combustion process and pollutant performance for ammonia-diesel dual-fuel engines. Energy 296

Shin J, Park S (2024) Numerical analysis and optimization of combustion and emissions in an ammonia-diesel dual-fuel engine using an ammonia direct injection strategy. Energy 289

Singh AP, Kumar D, Agarwal AK (2021) Alternative fuels and advanced combustion techniques as sustainable solutions for internal combustion engines. Springer Nature

Soldatkina V (2023) Fuel the EU: decarbonising European shipping

Sorenson SC (2001) Dimethyl ether in diesel engines: progress and perspectives. J Eng Gas Turbines Power 123(3):652–658

Tian J et al (2024) Visualization study on ammonia/diesel dual direct injection combustion characteristics and interaction between sprays. Energy Convers Manag 299

Tornatore C et al (2022) Ammonia as green fuel in internal combustion engines: state-of-the-art and future perspectives. Front Mech Eng 8:944201

Tu H, Liu Z, Zhang Y (2024) Study on cost-effective performance of alternative fuels and energy efficiency measures for shipping decarbonization. J Mar Sci Eng 12:21

Valera-Medina A et al (2018) Ammonia for power. Prog Energy Combust Sci 69:63–102

Wang D et al (2021) Numerical study of the premixed ammonia-hydrogen combustion under engine-relevant conditions. Int J Hydrogen Energy 46(2):2667–2683

Wang X et al (2024) Exploring the GHG reduction potential of pilot diesel-ignited ammonia engines—effects of diesel injection timing and ammonia energetic ratio. Appl Energy 357

Westlye FR, Ivarsson A, Schramm J (2013) Experimental investigation of nitrogen based emissions from an ammonia fueled SI-engine. Fuel 111:239–247

Xin G et al (2022) Effect of different volume fractions of ammonia on the combustion and emission characteristics of the hydrogen-fueled engine. Int J Hydrogen Energy 47(36):16297–16308

Xu X et al (2022) Review of the current status of ammonia-blended hydrogen fuel engine development. Energies 15(3):1023

Yapicioglu A, Dincer I (2018) Performance assesment of hydrogen and ammonia combustion with various fuels for power generators. Int J Hydrogen Energy 43(45):21037–21048

Yousefi A et al (2022) Effects of ammonia energy fraction and diesel injection timing on combustion and emissions of an ammonia/diesel dual-fuel engine. Fuel 314:122723

Zhang Z et al (2024) Effect of liquid ammonia HPDI strategies on combustion characteristics and emission formation of ammonia-diesel dual-fuel heavy-duty engines. Fuel 367

Zheng L et al (2024) Experimental study on the combustion and emission characteristics of ammonia-diesel dual fuel engine under high ammonia energy ratio conditions. J Energy Inst 114

Zhou X et al (2023) Pilot diesel-ignited ammonia dual fuel low-speed marine engines: a comparative analysis of ammonia premixed and high-pressure spray combustion modes with CFD simulation. Renew Sustain Energy Rev 173

Zhu D, Shu B (2023) Recent progress on combustion characteristics of ammonia-based fuel blends and their potential in internal combustion engines. Int J Automot Manuf Mater

Chapter 3
Design of Prechamber Jet Ignition System for Lean Combustion of Fuels

Rahul Kumar Singh and Avinash Kumar Agarwal

Abstract Prechamber is an advanced ignition technology used in spark ignition engine combustion. The prechamber ignition system allows the engine to operate for a lean air–fuel mixture by enhancing turbulence by an order of magnitude. The objective of using this technology is to achieve low emission and high engine brake thermal efficiency. The two types of prechambers are active and passive. An active prechamber can achieve ultra-lean combustion of fuels with high brake thermal efficiency. A passive prechamber operates at a lean limit slightly better than a conventional spark ignition engine. However, the advantages of the passive prechamber are high knock resistance, short combustion duration, and combustion stability. The objectives of prechamber design are to achieve fast flame propagation in the prechamber, proper scavenging, and high turbulence in the main chamber. Also, the prechamber is designed so that minimum modifications in the engine are made, and it can be easily replaced with a spark plug. The different design parameters of the prechamber are prechamber volume, the shape of the prechamber, nozzle diameter, the number of orifices, and the diameter of the orifice. The current chapter gives an introduction and history of prechamber concept development. This chapter reviews the state of the art of designing different parameters of the prechamber. The physical phenomenon occurring inside the prechamber and its importance in efficient design have been discussed. The effect of different prechamber parameters on physical phenomena in the prechamber, combustion, emissions, and performance of an engine, such as flame front propagation, charge motion in the prechamber, jet ejection, and scavenging processes, in-cylinder pressure, heat release rate are discussed. This chapter also discusses constant-volume combustion chambers, rapid compression machines (RCM), and engine studies conducted for the design optimization of prechambers. A summary of normalized optimum prechamber parameters based on the literature survey range is tabulated, which can be used by the readers as a reference for their design. Overall, this chapter gives an overview of the design of the prechamber and suggests the identified research gaps.

R. K. Singh · A. K. Agarwal (✉)
Engine Research Laboratory, Department of Mechanical Engineering, Indian Institute of Technology Kanpur, Kanpur, India
e-mail: akag@iitk.ac.in

© The Author(s), under exclusive license to Springer Nature Singapore Pte Ltd. 2026

H. Valera and A. K. Agarwal (eds.), *Ammonia and Low Carbon Combustion Engine Technologies for De-Fossilizing Transport Sector*, Energy, Environment, and Sustainability, https://doi.org/10.1007/978-981-96-6628-7_3

Keywords Turbulent jet ignition · Active prechamber · Passive prechamber · Prechamber design · Lean combustion

3.1 Introduction

The tailpipe emissions from vehicles are always a concern due to their effect on the environment, such as the increase in greenhouse gases (GHGs), which are responsible for climate change and human health. Zero and low carbon emissions from internal combustion (IC) engines are the goals to be achieved as soon as possible. Another challenge is to improve the thermal efficiency of the IC engines. Several technologies have been developed to overcome these two challenges and improve IC engine vehicles to continue as the backbone of transportation. Prechamber (PC) jet ignition systems are promising technologies for spark ignition (SI) engines that can help achieve high thermal efficiency and low tailpipe emissions (Gussak 1975). This technology is under development and has challenges that need to be tackled so that it can be used in vehicles. In this chapter, the discussion is on the development of the design for the PC jet ignition system.

3.1.1 Lean Combustion of Fuels

Lean combustion occurs when the actual air-to-fuel ratio is more than the stoichiometric, i.e., $\lambda > 1$. The combustion characteristics of lean air–fuel mixtures are different from those of rich and stoichiometric air–fuel mixtures. It produces less emissions and increases combustion efficiency. The excess air available during combustion for a lean mixture burns the fuel completely. The combustion temperature is lower for lean mixtures than for stoichiometric ones, which limits the formation of NOx. Another advantage of lean combustion is that for a given octane number, a higher compression ratio engine can operate without preignition.

The SI engines operate on a stoichiometric air–fuel mixture because the three-way catalytic (TWC) converter efficiency is maximum, and it reduces NOx, carbon monoxide (CO) and hydrocarbon (HC). Also, a slightly richer mixture than stoichiometric gives maximum power (Heywood 1988). However, stoichiometric operation restricts the SI engine from achieving maximum thermal efficiency. The reasons are high heat transfer due to high combustion temperature, especially for lower values of the ratio of specific heat capacity, which is attributed to the high percentage of triatomic gases. High combustion temperature also reduces the ratio of specific heat, which reduces the work extraction during the expansion stroke. Lastly, the dissociation of CO_2 and O_2 for a stoichiometric mixture inhibits the completion of combustion near the top dead centre (TDC). Lean combustion can overcome these limitations of stoichiometric combustion. However, the cyclic variation increases when the mixture is lean in an SI engine, which affects the drivability of the vehicle.

The reducing agents such as CO and HC are low for lean mixture combustion in an SI engine, which is why the performance of the three-way catalytic converter is very poor. The other challenge associated with lean mixture combustion of low flammability range of fuels, it will result in engine misfires and the cyclic variations will be higher. The oxidation rates of CO and HC formed in the crevices are lower for lean combustion due to a decrease in combustion temperature. Incomplete combustion or misfire for multiple cycles is frequent for lean combustion in an SI engine, which increases the HC emissions (Heywood 1988). The engine misfire occurs due to the absence of the lowest limit of air–fuel mixture around the flame kernel, which can be ignited. Another reason for the engine misfire is the slow flame speed which limits the growth of the initial flame kernel. The HC increases due to incomplete combustion and misfire. The development of suitable combustion technology for lean mixture combustion in an SI engine has been attempted to utilize the advantage of lean mixture combustion while overcoming the limitations. Some of the technologies developed are the laser ignition system, the use of a sparkplug with high ignition energy and increasing turbulence, and the PC jet ignition system (Heywood 1988; Srivastava and Agarwal 2013; Liu and Zheng 2024; Jung et al. 2017; Toulson et al. 2010). The laser ignition system is very complex. It is very costly, needs very precise control, and is difficult to install. For spark plug technology, increasing the spark discharge energy and discharge time is also a known technology that increases the efficiency of lean combustion within an acceptable coefficient of variation (COV). The flame kernel grows rapidly into a turbulent flame front and travels throughout the combustion chamber. There is a high probability that the flame kernel extinguishes early due to low turbulence intensity. Therefore, enhancement in the turbulence intensity is also essential for efficient lean combustion in an SI engine (Jung et al. 2017). Another technology for lean burning in an SI engine is the PC jet ignition system. In this system, a small quantity of fuel burns in the PC, which is injected at high pressure into the main chamber (MC) in the form of turbulent jets. The ignition energy of these jets is higher. The combustion products coming out of the PC are classified into two categories: flame ignition and jet ignition (Biswas 2018). In the flame ignition, the flame front does not get quenched in the orifice while moving towards the MC, whereas in the jet ignition, the flame front gets quenched in the orifice. The jet ignition mode has chemically active species or combustion products of the quenched flame, which reignite in the MC. There are many advantages of the PC jet ignition system. It can increase the flammability limit of the charge in the MC, enhance combustion, and lower cyclic variations and NOx emissions (Alvarez et al. 2018). This chapter discusses the development of the PC jet ignition system for lean or ultra-lean combustion. The next section discusses the history of the development of the PC ignition concept.

3.1.2 History of Prechambers

PC combustion in internal combustion engines has a long history of over a century. The first SI engine with a passive PC ignition system was employed in the Ricardo Dolphin Engine in the 1910s (Turkish 1974). Rich fuel–air mixture from the auxiliary inlet valve enters the PC cavity. The rich mixture was ignited by the spark, which then ignited the lean mixture in the MC. Ricardo Dolphin's engine design inspired the development of several more designs of PC ignition systems. The modern PC design has an auxiliary fuel injector that supplies the fuel into the PC. The modern PC is divided from the main combustion chamber by an orifice. This allows a stratified air–fuel mixture to form in the PC, which ignites a lean mixture in the MC. It was also called a divided PC. In the early days, the PC volume and orifice size were large, so a flame front propagates gradually from the PC to the MC via an orifice, which ignites the air–fuel mixture in the MC. The jet ignition concept was developed by Nikolai Semenov in the 1950s (Nikolai 1990). In 1974, Davis used the jet ignition concept and developed an ignition named Jet Ignition Stratified Charge Engine (JISCE). In 1975, Lev Ivanovich Gussak used this concept to develop the first engine with a jet ignition system named Lavinia Aktivatisia Gorenia (LAG) (Gussak et al. 1979). Gussak proposed the LAG design as the optimized design for the jet ignition system. The volume of the PC and orifice is very small compared to the old PCs, such as the torch cell ignition system. The small orifice helps to accelerate the burning mixture from the PC to the MC. The flame during the ejection from the orifice gets extinguished due to the quenching effect. These reacting species ignite the fuel–air mixture in the main combustion chamber. The flame jet penetrates much deeper into the MC, and wall impingement is avoided by making the PC volume small (Gussak et al. 1979). The combustion duration for the jet ignition system is very short compared to the spark ignition system.

In the period between 1968 and 1972, Honda developed the Compound Vortex Controlled Combustion (CVCC) ignition system. It could meet the emission norms in 1975 without a catalytic converter (Gussak et al. 1983). In 1975, Porsche developed a Stable Kernel of combustion (SKS) ignition system. The PC used a rich mixture to lean mixture ($\lambda = 1.5$ to 3) in the MC. Only the HC emissions were beyond the limit for SKS, whereas other emissions were under the limit. In 1981, using the LAG jet ignition system design, combustion of ultra-lean ($\lambda > 2$) was achieved by fuelling a rich mixture ($\lambda = 0.5$) into the prechamber (Alvarez et al. 2018). In 1979, Adams studied the torch jet ignition system (Adams 1979). This system was developed by multiple manufacturers like Toyota, Ford, and Volkswagen (Alvarez et al. 2018; Adams 1979; Noguchi et al. 1976; Brandstetter 1980). In this system, the auxiliary inlet valve was replaced by a spark plug. The PC was fuelled by the fresh charge in the MC during the compression stroke. After spark ignition, a turbulent torch ejects from the PC, which ignites the MC charge (Toulson et al. 2010). In the 1990s, Hydrogen Jet Assisted Ignition (HAJI) system was developed at the University of Melbourne to ignite an ultra-lean mixture (Lumsden and Watson 1995). The lean mixture limits up to $\lambda = 5$ at wide open throttle (WOT) with gasoline as MC fuel and hydrogen as PC fuel can be burned with HAJI. HAJI uses active chemical species and turbulent jets ejected from the PC to ignite the MC charge. The MC was supplied with gasoline fuel, and hydrogen was supplied to the PC. The PC energy was between 1 to 3% of the MC fuel. The fuel consumption and emissions were reduced for $\lambda < 2$ (Lumsden and Watson 1995). The HC emissions increased significantly for λ between 2 and 5 due to low temperature and wall quenching. Robinet et al. (Quader 1974) developed a Self-Ignition Triggered by a Radical Injection (APIR) system in 1999, which has a volume of 1% of the clearance volume. This system shows an improvement in fuel consumption at low load. In 1984, Latsch developed a swirl chamber with a spark plug, which further improved fuel consumption and exhaust emissions. The Bowl-Prechamber-Ignition (BPI), which utilizes the concept of homogeneous charge ignition. A double injection strategy is used for this ignition system. The first injection is done during the intake stroke, and the second injection is done during the compression stroke to ignite the mixture. It is a kind of low-temperature combustion strategy. In 2012, Toulson et al. (2012) developed a turbulent jet ignition (TJI) system by replacing the spark plug with a PC, which has a fuel injector and spark plug. This ignition system showed many advantages, such as short combustion duration and combustion of ultra-lean mixture ($\lambda = 1.8$). After the development of the TJI system, many PC designs have been developed for the PC jet ignition system in the past two decades. In the next section, the discussion will be on the classification of PCs.

3.2 Types of Prechambers

In the previous section, the discussion highlights the development of several PC jet ignition systems over a century. Based on the fuelling of the PC, these PCs can be classified into two categories: passive and active PC. Figure 1a, b represent the two PC categories. These two types of PCs are discussed in the next sub-sections.

3.2.1 Prechamber

In a passive PC ignition system, the fuel is supplied to the PC from the MC during the compression stroke via the orifice of the PC. Figure 1a represents the passive PC configuration. The spark plug is attached to the passive PC chamber. There is no auxiliary fuel injector attached to the passive PC. The MC is fuelled by a port fuel injector (PFI) or a direct injector (DI) injection system. The spark ignites the mixture in the PC. The flame front propagates from the PC toward the MC via an orifice. The flame gets quenched at the orifice due to its small size. The turbulent jets coming out from the orifice contain reacting species. There can be single or multiple jets depending on the number of orifices. The turbulent jet acts as a multipoint ignition source, which ignites the mixture in the MC. The composition of the mixture in the

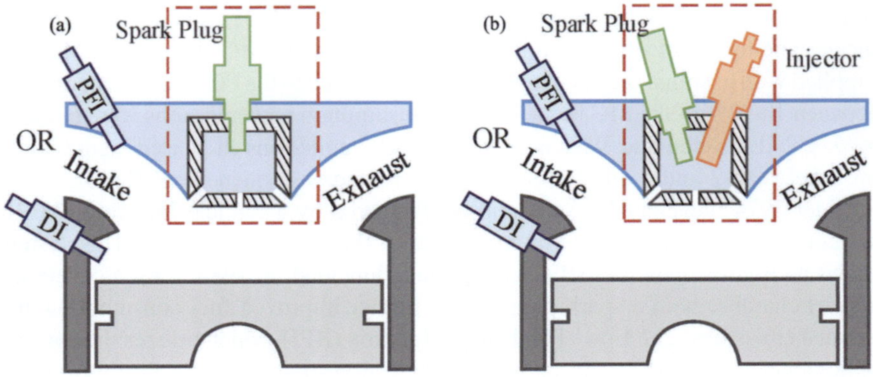

Fig. 3.1 **a** Passive PC and **b** active PC configurations (Zhu et al. 2022)

PC can't be controlled differently from the MC. Thus, the composition of the fuel–air mixture in the PC and MC is the same. That is why these types of PCs are also called homogenous PCs. The fresh fuel–air mixture must enter from the same orifice into the PC, which causes a substantial portion of the burnt gases to remain in the PC. That is why a passive PC is also known as a unscavenged PC. The passive PC is simpler, cost-effective, and easy to fit in place of the spark plug. The limitation of passive PC is the fresh air–fuel mixture filling into the PC and the poor scavenging process (Blankmeister et al. 2018; Benajes et al. 2020). During the intake stroke, the filling of PC is only 15% of the total volume, which is much less (Benajes et al. 2020). The gas flow processes in the passive PC are discussed in detail in Sect. 3.2 of this chapter. The lean limit of the engine with a passive PC is like the SI engine. The advantage is that the combustion duration is very short, which is why the thermal efficiency is higher with low cyclic variations.

3.2.2 *Active Prechamber*

In an active PC ignition system, the fuel is supplied to the PC via an auxiliary fuel injector attached to the PC. Figure 1b represents the configuration of an active PC. A spark plug is also attached to the active PC, like the passive PC, which ignites the mixture in the PC. The fuel–air mixture in the MC is lean. During the compression stroke, the fuel–air mixture from the MC enters the PC. A small quantity of fuel is injected into the PC using the auxiliary injector to make the fuel–air mixture composition rich in the PC. It resembles the stratified mixture formation in the gasoline direct injection engine. The advantage of this kind of PC configuration is that a rich air–fuel mixture can be formed in the vicinity of the spark plug. Thus, it is also known as stratified PC. In an active PC, especially for gaseous fuel, the fuel injection assists the scavenging process. Additional scavenging methods may be required for very lean engine operation for active PCs. The two methods tested by the researchers are air injection into the PC and injection of fuel in the MC and PC using a single-fuel injector (Vedula et al. 2017; Krajnović et al. 2024). The quality of the mixture formed in the PC can be controlled precisely due to the presence of an auxiliary fuel injector. The PC and MC charge compositions are different for the active PC configuration. Usually, a fuel–air mixture of the MC is very lean, and the PC is rich for an active PC. The process in the active PC is distinct from the passive PC. The active PC is more efficiently scavenged compared to the passive PC. That is why it is sometimes called a scavenged PC. The lean operation limit for an active PC engine is very high compared to an SI engine. This ignition system can go up to $\lambda = 2.2$, and this lean mixture is termed the ultra-lean fuel–air mixture. The gas exchange process between MC and PC in an active PC is different than that in a passive PC. The details of the gas flow in both active and passive PCs are discussed in Sect. 3.2 of this chapter.

3.3 Gas Flow in Prechamber

In this chapter, all the gas flow in the PC system is classified into four phases. These are fuel–air filling, charge-ignition in PC, jet ejection, and scavenging process. These processes are discussed in detail in the next subsections. Elia et al. (2020) classified the process in an active PC configuration into different phases, as shown in Fig. 3.2. These processes are filling and scavenging, mixing, flame propagation, ejection, burning, and expulsion and extraction. More details on the processes can be found in the cited reference.

3.3.1 Charge Filling

The fuel–air mixture filling process in the active and passive PC is distinct. In the active PC, during the compression stroke, lean fuel–air enters the PC, and an auxiliary fuel injector supplies the fuel so that the fuel–air mixture composition near the spark plug is rich. Chinnathambi et al. (2015) conducted a Reynolds-averaged numerical simulation (RANS) to study the active PC fuel-filling process, and they found that the fuel spray interacts with the crossflow of the incoming charge from the MC and impinges on the wall of the PC.

The design of the PC chamber for the incoming charge plays a key role in scooping the fuel impinged onto the wall of the PC and transporting it to the other region. The impingement of the spray onto the wall is unavoidable in an active PC, as shown in Fig. 3.3. The spray, after hitting the wall of the PC, creates a tumble motion, which enhances the mixing, as shown in Fig. 3.3 (Zhao et al. 2024). As the compression stroke continues, a stratified mixture is formed near the spark plug. The velocity of the charge near the spark plug is higher than that in the conventional SI engine. The fuel–air mixing in the PC is affected by the PC geometry and orifice arrangement (Chinnathambi et al. 2015). On the other hand, the fuel–air mixture enters the passive

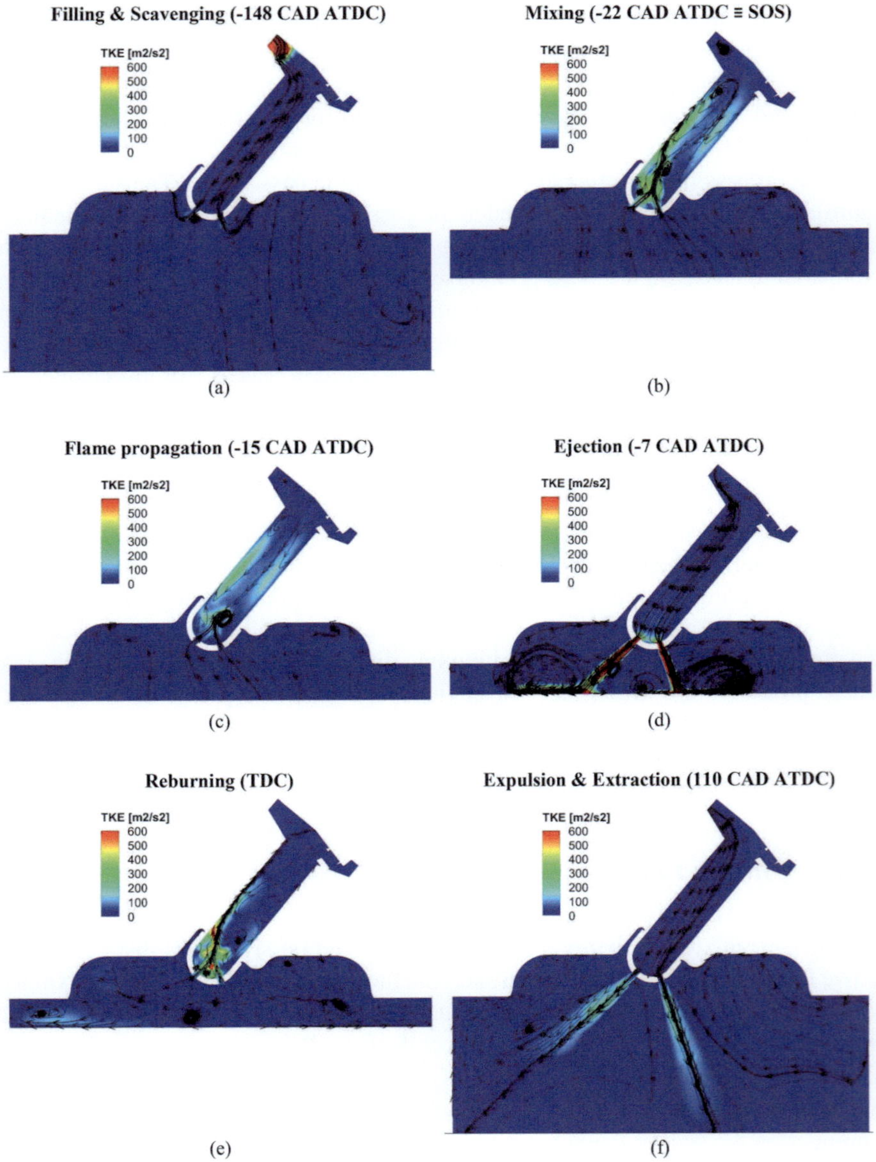

Fig. 3.2 Evolution of different processes for an active PC jet ignition system (Distaso et al. 2020)

PC during the compression stroke. The mixture composition in the MC and PC is the same since no auxiliary fuel injector is used in this configuration. The quantity of fuel–air mass filled in the passive PC is affected by the engine speed and load. The turbulence and mixture stratification in the passive PC play a key role in the efficient combustion of charge in the PC (Novella et al. 2022). The turbulent kinetic energy

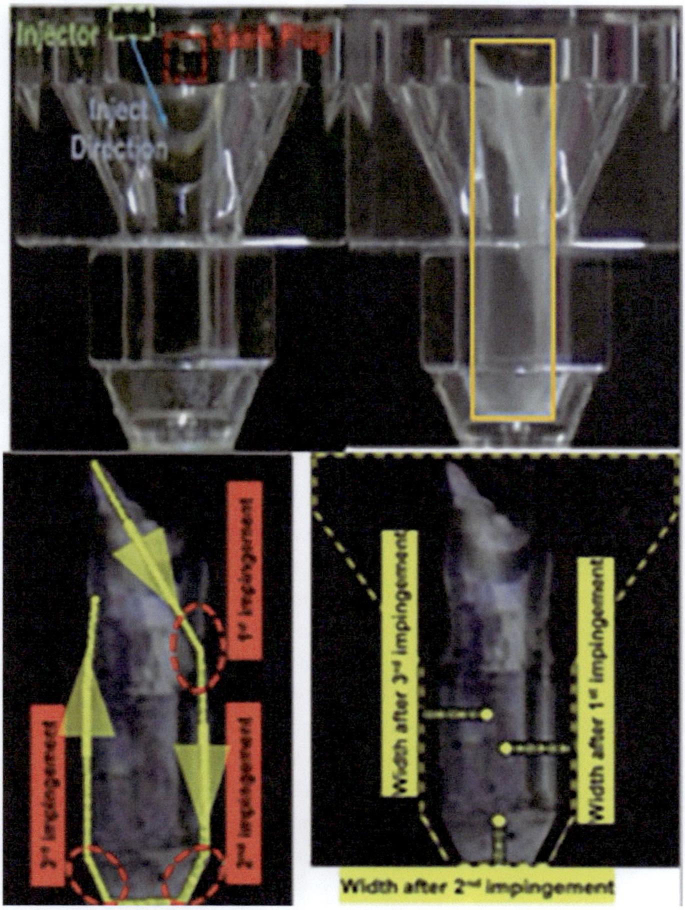

Fig. 3.3 Tumble flow of fuel spray inside an active PC (Zhao et al. 2024)

(TKE) of the charge in the PC is increased by the fuel injected in the active PC and the charge inflow from the MC to the PC (Distaso et al. 2020). The increase in TKE by the flow near the spark may harm the initial flame growth (Ho and Santavicca 1987). Therefore, the increase of TKE near the spark plug should not be significant.

3.3.2 *Ignition*

The spark in the PC initiates the combustion of the air–fuel mixture. The combustion in the PC is a combined interaction of fuel composition, flow dynamics, and reaction chemistry (Zhao et al. 2023). The initial phase of combustion in a PC is like that of

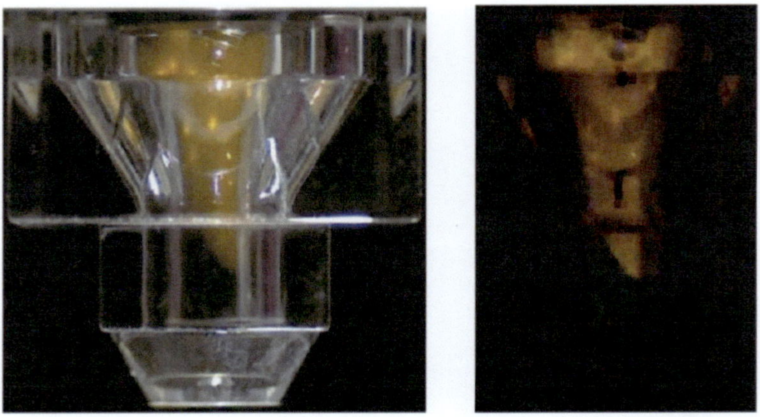

Fig. 3.4 Ignition and flame propagation process in a transparent active PC (Zhao et al. 2024)

a conventional SI engine. The differences between these two are that the surface-to-volume ratio and turbulence are higher for PC (Vedula et al. 2017). The initial flame kernel in the small PC burns most of the charge. The flame front propagates towards the PC orifices. Another fact is that the high turbulence inside the PC is not suitable for combustion in the PC. The small volume of PC laminarizes the flow near the wall, which reduces the flame speed. The size of the PC limits the turbulent integral length scale in the PC. The length scale in the MC is large. Thus, engine simulation becomes a bit challenging for a PC ignition system due to the wide range of integral length scales. Gussak et al. (1975) concluded that the rich fuel–air mixture in the active PC results in more active radicals up to a certain limit.

Figure 3.4 shows the ignition and flame propagation processes in a transparent active PC (Zhao et al. 2024).

3.3.3 Jet Ejection

The essential parameters that influence the jet ejection from the PC are the pressure difference between the PC and MC, and jet momentum (Novella et al. 2022). A high-pressure difference and jet momentum are preferred for quick jet ejection. It also depends on the thermochemical conditions in the PC (Benajes et al. 2020). The local pressure increase in the PC forces the flame front into the orifices. The flame front gets quenched in the orifice and converted to a jet of combustion products or active species (Idicheria et al. 2018). The thermal quenching in the orifice is due to the heat loss to the wall of the orifice, which is far greater than the heat generated due to the chemical reaction. Kyrtatos et al. (2018) explained that due to quenching, the chain-branching reaction inhibits only the radical recombination reaction that takes place in the orifice. Thermal quenching occurs in the orifice. The ejection of flux from the PC is in three stages.

i. During the initial discharge process of the jet, a significant amount of fuel–air mixture is pushed by the flame front from the PC to the MC via orifices. However, these fuel–air mixture jets are consumed by the high-momentum ignition jets behind them. It results in a localized high-temperature zone in the MC.
ii. Then, the hot jet ejects from the PC orifice to the MC. These jets in the MC function as multiple ignition sites that engulf the lean mixture in the MC rapidly (Chinnathambi et al. 2015).
iii. In the last stage, after the ejection of the high-velocity species-rich jet from the PC to the MC, there is a reverse flow of the combustion gases from the MC into the PC. Due to reverse flow, reburning in the PC occurs. Reburning accelerates the oxidation of the partial combustion products in the PC and enhances species diffusion (Distaso et al. 2020). Hydrodynamic quenching occurs in the MC due to the intense mixing of unburnt charge and the combustion products in the jet. Sometimes, this may result in engine misfires due to global quenching in the MC (Zhu et al. 2022).

For a passive PC, the jet ejection process is fast for high engine speed and load due to the large pressure difference between PC and MC and high jet momentum, as shown in Fig. 3.5 (Novella et al. 2022). In Fig. 3.5, the terms OP1, OP2, and OP3 represent three engine operating points, which are high, medium, and low loads and speeds, respectively. The jet momentum is ten times lower for low speed and load compared to high speed and load. The flow characteristics inside the PC deteriorate for low engine speed and load. The average TKE is higher for high engine speed and load compared to medium and low load and speed. It favours quick flame propagation at high engine speed and load (Novella et al. 2022). Thus, it is essential to optimize the passive PC design for the complete range of engine load and speed. Dilution of the air–fuel mixture in a passive PC engine reduces the pressure difference between MC and PC, as well as the momentum of the jet. Thus, the jet penetration into the MC is reduced, which increases the ignition delay and slows down the combustion in the MC (Frasci et al. 2023). A high level of TKE and temperature of the jet sustain the flame and consume the entraining charge in the MC (Benajes et al. 2020).

3.3.4 Scavenging

Scavenging is the process of pushing out the burnt gases by the fresh air–fuel mixture in the PC. During this process, a part of the fuel is also lost to the MC. The scavenging process of a PC is essential for stable engine operation. The passive PC has poor scavenging compared to the active PC configuration. The filling of the PC and scavenging process occur simultaneously in the active PC, and fresh charge assists the scavenging in the active PC. The gas motion in the PC is dependent on the crank angle due to the mass transfer between the MC and PC. Thus, geometric design and

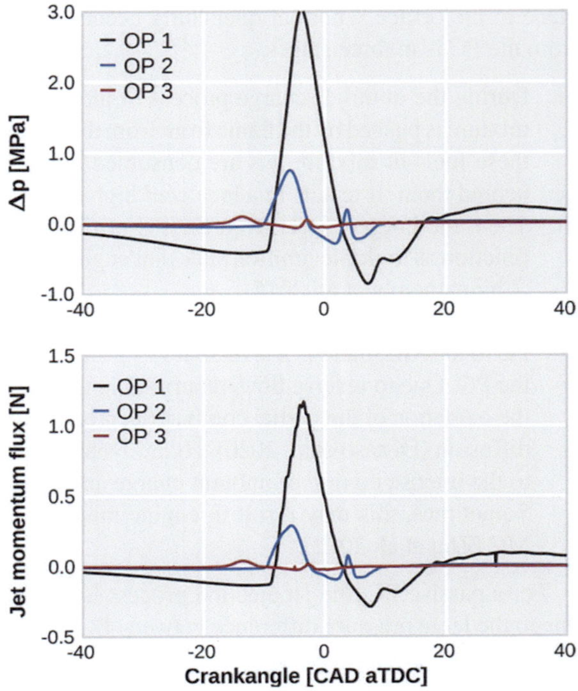

Fig. 3.5 Pressure difference and jet momentum for passive PC at different loads and speeds (Novella et al. 2022)

operating conditions affect the scavenging of the PC. For a lean air–fuel mixture with a passive PC, the spark timing must be retarded. However, the air–fuel mixing time available becomes short, and it results in a high concentration of combustion products in the PC.

3.4 Design Parameters of Prechamber

The design of the PC is very crucial for efficient and stable engine operation. The main design parameters common to passive and active PCs are the volume and shape of the PC, number of orifices (ON), size of orifice, which is orifice diameter (OD), and angle between, which is orifice angle (OA), as shown in Fig. 3.6 (Baumgartner et al. 2017). The design of an active PC is more complex than a passive PC. The additional parameter in the active PC configuration is in the position of the spark plug and fuel injector. The main objective of PC design is to control the thermal, chemical, and turbulence of the jet. The component position in the PC is responsible for the symmetry of the jet pattern in the MC (Chinnathambi et al. 2015). To ensure the easy retrofitment of the PC assembly in the SI engine, minimum modification is also a design factor that should be accounted for. The details of the PC design parameters are introduced in the next subsections.

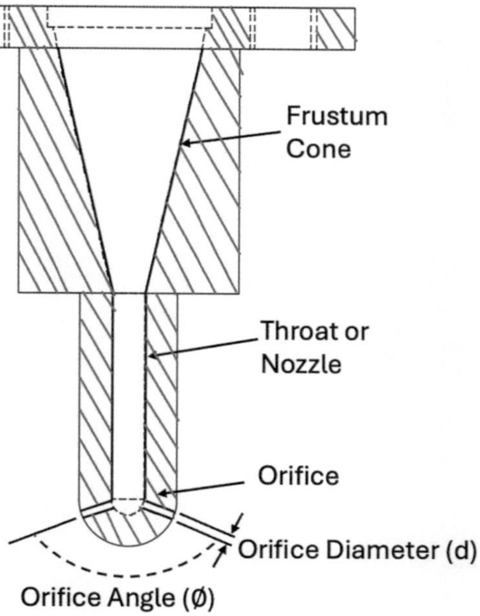

Fig. 3.6 Essential parts and parameters of PC

3.4.1 Prechamber Volume and Shape

PC volume governs the amount of mixture energy in the PC that can be transferred to the MC for ignition. A small volume of PC led to stable and efficient combustion. The amount of fuel inside the PC does not affect the power output of the engine. Thus, the minimum amount of fuel trapped in the PC, which can ignite the MC mixture, is between 2 and 4% of the clearance volume of the engine cylinder. The volume of the PC influences the engine's effective compression ratio (CR). The common shape, considering the space constraints of a PC, is in two parts. The top part is the frustum of a cone or a cylinder, and the bottom part is a nozzle. The bottom part, which is a nozzle, fits in the place of the spark plug. This design is also usually adopted for active PC configuration, Fig. 3.7 (Tang et al. 2021; Zhou et al. 2023). The large upper part of the active PC results in proper air–fuel mixing and turbulence (average TKE) (Serrano et al. 2019).

The design of the PC on the left in Fig. 3.7 is the narrow-throat PC design by Clean Combustion Research Centre, KAUST, Saudi Arabia. On the right side of Fig. 3.7 is the active PC assembly from the State Key Laboratory of Engines, Tianjin University, China. The arrangement of the injector and the spark plug in a Y shape is mostly adopted in the design of an active PC. The spray wall impingement in active PC is dependent on the spray angle. If the spray angle is inappropriate, a large amount of spray hits the PC wall (Koga et al. 2019). It depends on the injector location in the PC body and the angle of inclination. PC orientation affects the jet ejection from

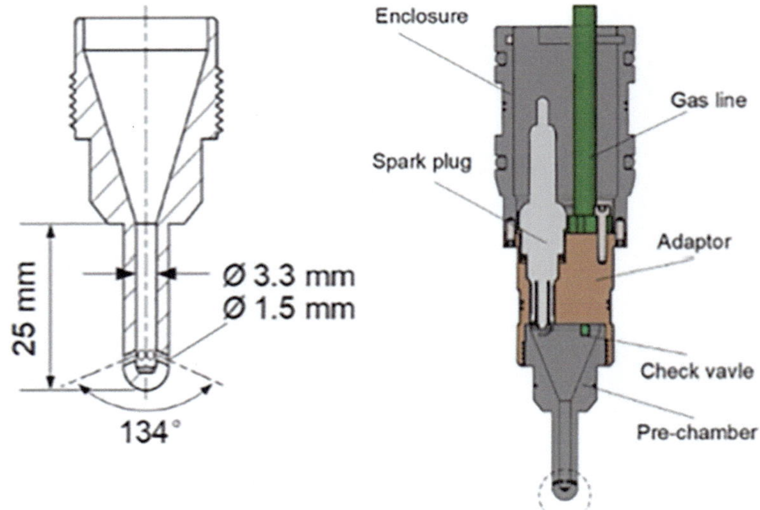

Fig. 3.7 The shape of active PC configuration (Tang et al. 2021; Zhou et al. 2023)

multiple orifices; the jet ejection from some orifices is quicker than the other (Distaso et al. 2020).

The passive PC mostly has a large nozzle (bottom part of the active PC) compared to the frustum cone, unlike the active PC, as shown in Fig. 8b (Antolini et al. 2023). The outer diameter of the nozzle may be slightly less than the original spark plug diameter. In many passive PCs, the frustum cone is avoided, and the spark plug is directly inserted into the nozzle by threading or bolts, as shown in Fig. 8a (Wang et al. 2024). The length of the PC is constrained by the existing dimensions of the space for the spark plug and the volume of the PC. If the shape of the PC is cylindrical, then the length is usually equivalent to the length of the threads of the spark plug. However, it can be small or larger depending on the volume of the PC selected, based on the clearance volume of the engine. For an active PC, the length is larger than the threads for the spark plug due to the frustum of a cone on the top of the PC throat.

3.4.2 *Prechamber Orifices*

Another essential feature of the active and passive PCs is the orifices through which the products of partial combustion from the PC are transferred to the MC. The design parameters for the orifice are the number, diameter, and angle of the orifice (Zhu et al. 2022). A combination of different orifices with varying parameters is shown in Fig. 3.9. These have different nozzle numbers, cross-sectional areas, diameters, and orientations (Guo et al. 2024).

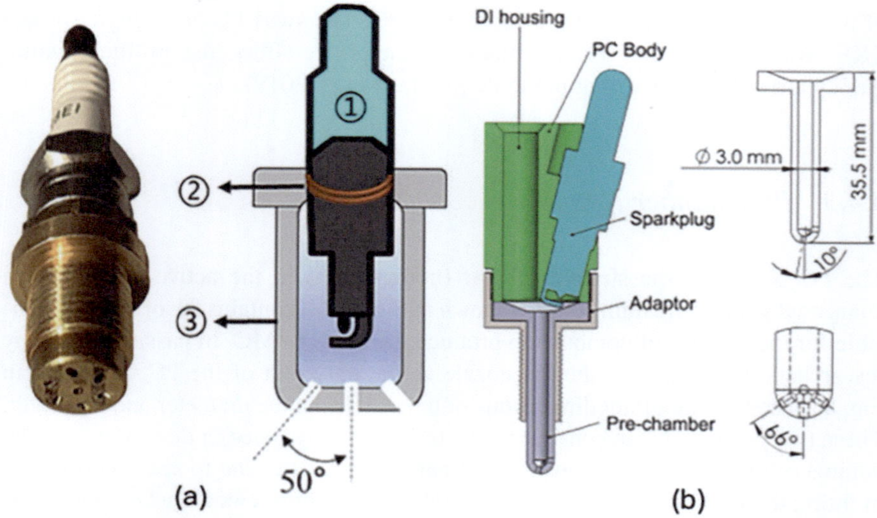

Fig. 3.8 Shapes of passive PC configuration (Antolini et al. 2023; Wang et al. 2024)

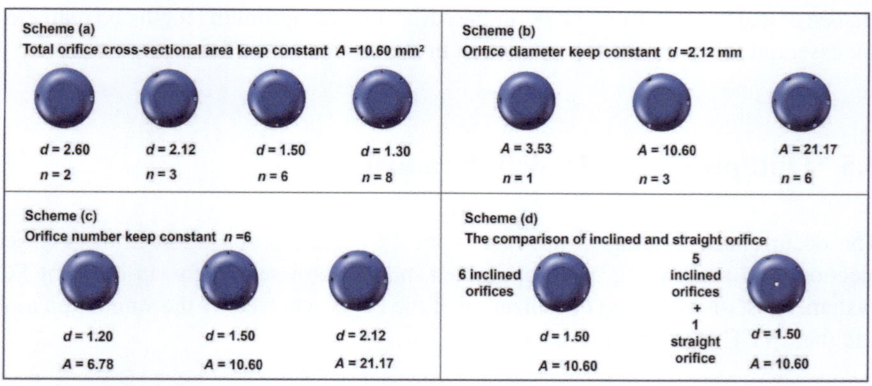

Fig. 3.9 Orifices with different diameters, numbers, and orientations (Guo et al. 2024)

There can be single or multiple orifices in a PC. The diameter of the orifice influences the scavenging, flame-quenching, and jet velocity. For a small orifice and a large PC volume, the mass of the burnt gases trapped will be higher. On the other hand, a large diameter will result in the loss of charge from PC to MC (Zhu et al. 2022). The flame quenching in the orifice is governed by its diameter. A large orifice will lead to torch ignition, whereas a small diameter will result in the jet ignition mode. These two modes have been discussed in the early part of this chapter. Too small a size of the orifice will result in inappropriate jet formation, which may extinguish without igniting the air–fuel mixture in the MC. The orifice angle between the orifice axis and the PC axis influences the turbulence and the scavenging process. The axis of the orifice and PC are skewed for a swirl PC. The swirl PC has a high average TKE (Wang et al. 2024). The orientation of the orifice affects the air–fuel mixture quality around the spark plug in the PC (Serrano et al. 2019).

3.4.3 Prechamber Nozzle

The PC Nozzle is the straight tubular (bottom part) in the active configuration connected with the frustum cone, as shown in Fig. 3.6. It contains the orifices responsible for the transfer of combustion products from PC to MC. In passive PC, many researchers designed only the PC nozzle as the only part of the PC, as shown in Fig. 8a. The two important dimensions of the nozzle are the diameter and its length. These two parameters are constrained by the spark plug space in the engine and the volume of the PC. KAUST developed a narrow-throat PC due to space constraints in fitting the PC assembly (Tang et al. 2021). A long throat would take a long time for the flame front to reach the orifice, whereas a short throat would take less. Due to this, spark timing must be advanced, and a suitable mixture should not be formed in the PC. A short throat may have a short flame front travel time, but there is a high chance of leaking fuel from the PC to the MC. Thus, an optimum length and diameter are essential for flame front propagation in the throat of the PC.

3.5 Optimization of Design Parameters

The occurrence of filling of air–fuel in the PC, flame propagation in the PC, jet ejection, and scavenging in an optimized manner is essential for an efficient PC system. Most of the design optimization of the PC is achieved by the simulation tool and the CVCC experimental setup.

3.5.1 Simulation Studies

Chinnathambi et al. (2015) investigated the active PC using a RANS-based multidimensional model. They found that the holes drilled for the instrumentation of the PC system are potential sites for autoignition. The location of the components in the PC results in the jet asymmetry in the MC. Benajes et al. (2020) found that for a constant PC volume to total orifice area ratio (V/A) in a suitable range, the scavenging process is primarily controlled by the piston momentum transferred to the air–fuel mixture during the compression stroke. Also, for a large V/A ratio, the scavenging process deteriorates. The residual gases in the PC are related to the compression ratio of the engine and cannot be evacuated from the PC. Thus, it acts as an internal exhaust gas recirculation (EGR). Benajes et al. (2019) optimized the PC volume and diameter by using 1D modelling. The authors found that a small volume pressurizes the PC significantly, but the energy in the PC is low for a small volume. A large diameter enhances the charge filling from MC to PC for a passive PC configuration. However, a large orifice results in the loss of charge back to MC. Liu et al. (2023) investigated five active PCs with different volumes and orifice diameters using 3D simulation. The two layers of the orifice were made on the PC. The author observed a cold jet ejection from the PC ahead of the hot jet. The PC designs A2 and B1 result in the ejection of a more intense reacting jet flame due to a small OD. These intense jets were ejected from the upper layer of the orifice due to the gas choking effect. The other observation was the influence of the PC geometry on the jet flame pattern and PC heat release rate (HRR). However, the MC HRR was not affected by the PC geometry much because HRR in the active PC during the later stage is governed by the flame front propagation towards the lean mixture in the MC. The author concluded with the recommendation of optimization of PC and piston geometry simultaneously (Fig. 3.10).

Silva et al. (2023) optimized the base design of the narrow throat PC developed by the KAUST research group (Tang et al. 2021) by design of experiments (DoE) and machine learning (ML) with the help of 3D simulations. A total of nine parameters, of which six independent and three dependent parameters, were used to optimise PC, keeping the PC volume constant (5.07 cc). The DoE was applied to 300 PC designs and the optimum one was selected based on the merit function. Further,

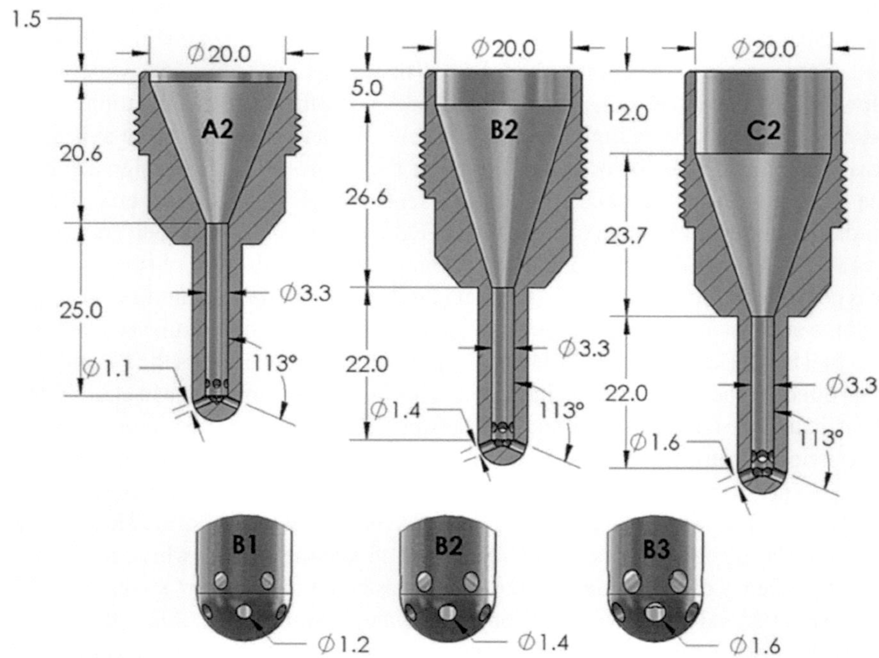

Fig. 3.10 Five active PC designs with different dimensions (Liu et al. 2023)

ML was implemented to obtain a globally optimized geometry (Silva et al. 2023). The best DoE design of the PC showed improvement in the baseline design of the PC, but compromised with NOx. The ML optimisation showed improvement for all the selected aspects. In another simulation study, Silva et al. (2022) investigated the effect of throat diameter on the combustion and turbulence characteristics of active PC systems. The large throat has a more stratified mixture around the spark plug due to the low flow speed of charge from the MC to the PC. The turbulent fluctuation component for the small throat diameter due to the large velocity causes better mixing in the PC. The integral length scale of the charge in the PC is bounded by the orifice diameter. Hence, the same was found for both throat diameters. The jet reaches the piston faster for a small throat diameter PC due to the higher-pressure difference between MC and PC than for a large throat diameter. The temperature of the burnt gases in the small throat is higher than in the large one due to less heat transfer.

Lv et al. (2024) investigated the effect of ON, orifice axis angle, OD, and volume ratio (VR) on combustion and emissions of passive PC engines using 3D simulation. The author reported that a reduction in the ON results in an enhancement of jet penetration and velocity. Another effect of ON reduction was a slight increase in the combustion duration (CD) in PC, which improved the jet flame velocity and decreased the CD in the MC. The emissions decreased as ON lowered from 6 to 5, except for NOx. The decrease in OD increased the flame jet velocity and the exhaust reflux in the PC. Three ODs were selected for the investigation. For the increase in OD, the soot, CO, and HC first increased and then decreased for the highest OD, but NOx followed opposite trends. A large VR achieved the highest temperature, and the probability of igniting the MC mixture was higher. This results in a short CD and advancement in the combustion phase. The soot and CO emissions were lower. However, NOx and HC emissions were high. Among the eight combinations of the passive PC, the PC with high OD and VR, low ON, and medium orifice axis angle was recommended as the optimum design of the passive PC. This design achieved the highest indicated thermal efficiency (ITE) (Lv et al. 2024). Frasci et al. (2023) studied the effect of the area of the orifice to PC volume (A/V) on the performance of passive PC using the 1-D model. A variation of 30% on both sides of the base value was considered for the study. The jet momentum increases with an increase in the PC volume at constant OD. The jet penetration reduces for a lower volume of PC due to less fuel in the PC, which results in a lower pressure difference across the PC and MC. This effect of the A/V ratio was independent of the air–fuel mixture. Hu et al. (2022) investigated the combustion differences between a symmetric and asymmetric PC for an aviation piston engine using simulations. The effect of OA on asymmetric patterns was also investigated. The results showed that the symmetric PC design performed better than the asymmetric PC. The cumulative heat release (CHR) was higher, and CD was shorter for asymmetric PC compared to symmetric PC, which resulted in better efficiency of the engine. An OA of 50° was better than 80° for asymmetric PC, which accelerated the flame growth and propagation. Muller et al. (2021) studied the active PC design parameters for different ON and orifice arrangements, as shown in Fig. 3.11. Two ONs, four and six, and the orifice arrangement were radial and at radial offset direction (swirl PC), were selected for the investigation. The swirl PC enhances

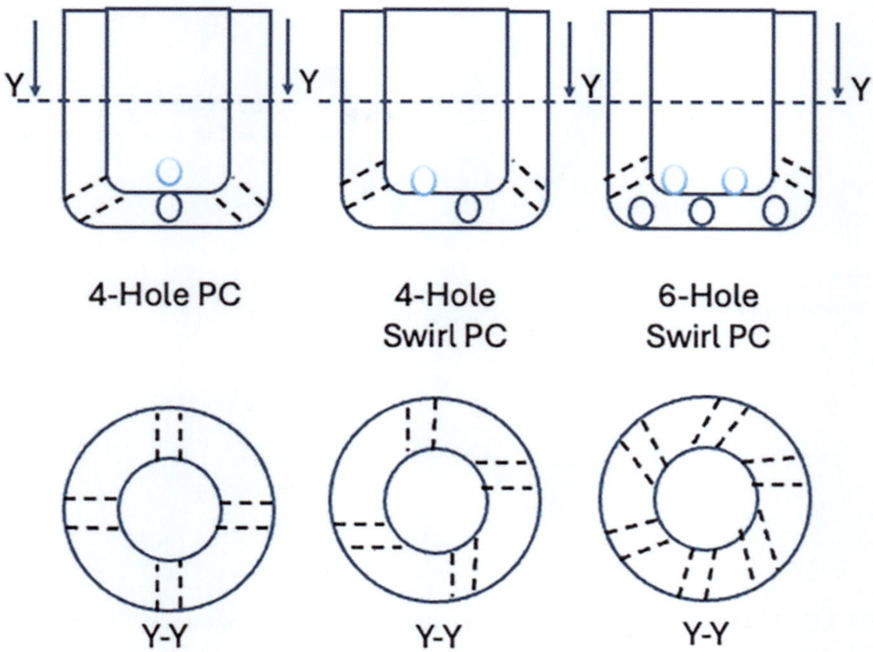

Fig. 3.11 PC design with different ON and orifice arrangements (Müller et al. 2021)

the turbulence and stratification of the mixture near the spark plug. A reduction in the HC emissions was observed for the jet hole orientation towards the quenching regions in the MC. The turbulence level of the 6-hole PC was higher compared to both 4-hole PCs, which reduced the ignition delay and maximum pressure rise rate in the PC. Soot formation was lower for the 6-hole PC than for the other two PCs due to high turbulence. However, the HC emissions were higher for 6-hole PCs than for 4-hole PCs.

3.5.2 CVCC and RCM Studies

An investigation of the design parameters of the processes of the active and passive chambers has been conducted in the CVCC and the Rapid Compression Machine (RCM). A setup developed by the research group at Michigan State University, USA, is shown in Fig. 3.12 (Gentz et al. 2017). The setup is attached to the active PC assembly, and its specifications are given in the cited reference. Gentz et al. (2017) investigated the effect of ON on the combustion characteristics. Single and double orifice nozzles were selected as ON parameters for both passive and active configurations. A dual-orifice active PC shows different combustion characteristics from a single-orifice active PC. The jets from the dual orifice can consume more of

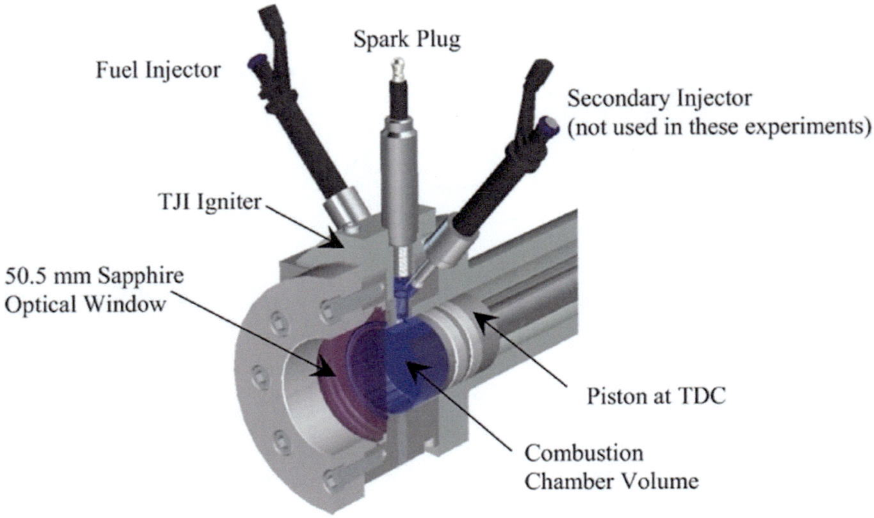

Fig. 3.12 RCM setup for the investigation of the parameters effect on processes in PC (Gentz et al. 2017)

the charge in the MC of RCM. The autoignition duration of the charge in the MC is shorter for a dual orifice than a single orifice for an active PC. The single orifice passive PC has better ignition characteristics than the dual orifice passive PC. It was evident from the 0–10% burn period and flame images as shown in Fig. 3.13 (Gentz et al. 2017). However, for an ignitable fuel–air mixture, both single and dual-orifice passive PCs are suitable for ignition of the charge in the MC.

Thelen et al. (2016) used the same experimental setup as shown in Fig. 3.12, and the simulation investigated the effect of spark location on the PC. Four points along the axis of the spark plug were selected. Based on 0–10% mass fraction burn, the author reported that the ignition location was away from the spark plug for the farthest spark location from the spark plug and produced the quickest ignition in the MC. However, the jet from the farthest ignition point was weak. On the other hand, the jet for the closest spark location from the spark plug produced jets with a high velocity, which penetrated the MC quickly and produced intense shear, which generated intense turbulence (Thelen and Toulson Apr. 2016). OD affects the 0–10% mass burn duration for the different air–fuel ratios, whereas it significantly affects the 10–90% mass burn duration in MC for the lean mixture. The smaller OD has the shortest 0–10% mass fraction burn duration for a passive PC due to the intense turbulence produced. Also, smaller ODs have a short 10–90% mass burn duration due to fast flame propagation in the MC (Thelen and Toulson 2016). Wang et al. (2022) designed nine PCs for a methanol-air mixture using the Taguchi method,

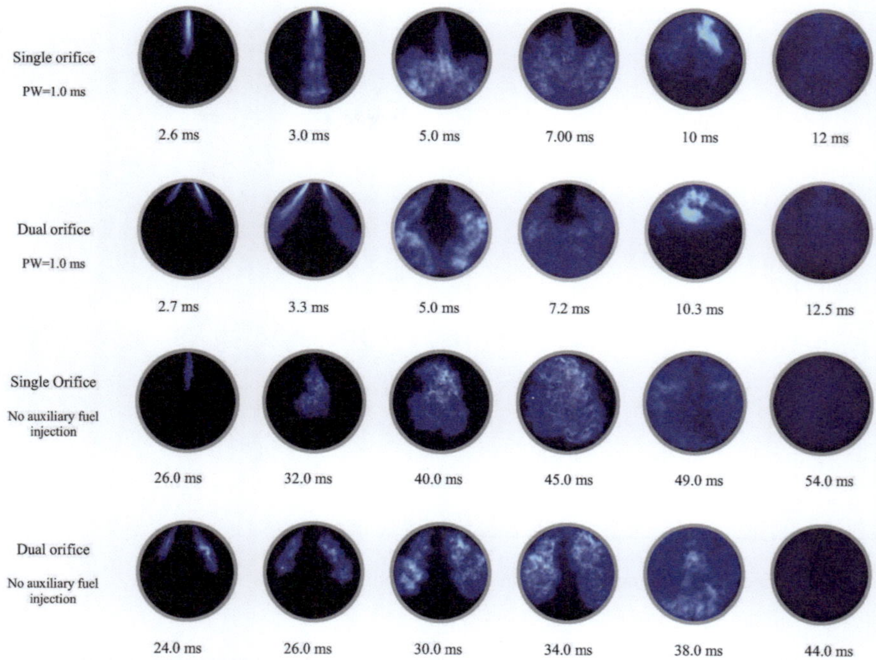

Fig. 3.13 Combustion images for single and dual orifices for active and passive PC configuration (Gentz et al. 2017)

as shown in Fig. 3.14. The study investigated the combustion characteristics of the methanol-air mixture in CVCC.

Four ignition modes were observed with nine PC designs. These ignition modes were jet flame ignition, jet ignition, jet wake ignition, and dual jet ignition. The HRR is more for four ignition modes compared to conventional spark ignition, with the highest increase for the dual-jet mode. The total area of the orifice affects the combustion of stoichiometric and lean air–fuel mixture. The jet wake and jet ignition mode can be achieved with a lower value of the total cross-sectional area of orifices. The authors also reported the effect of CVCC volume on the combustion characteristics, correlating it with ON. For large CVCC volume, four jets showed fast combustion, whereas for small CVCC volume, PC with six jets burned the MC charge more quickly. For a large number of jets, the OD becomes so small that the velocity of each jet is very low (Wang et al. 2022). A 60° transition angle (shapes 1, 2, and 3) gives a smooth movement of the combustion products from the PC top volume towards the orifice, and the pressure drop was minimal. A higher transition angle offered a large resistance to the flow of combustion products, which resulted in the jet penetration of the MC at a low speed. The authors also reported an insignificant effect of the throat diameter on the combustion process and recommended that it should be selected based on the space available and the volume of the PC. The most critical parameter that affected the lean-burn limit was the orifice cross-sectional

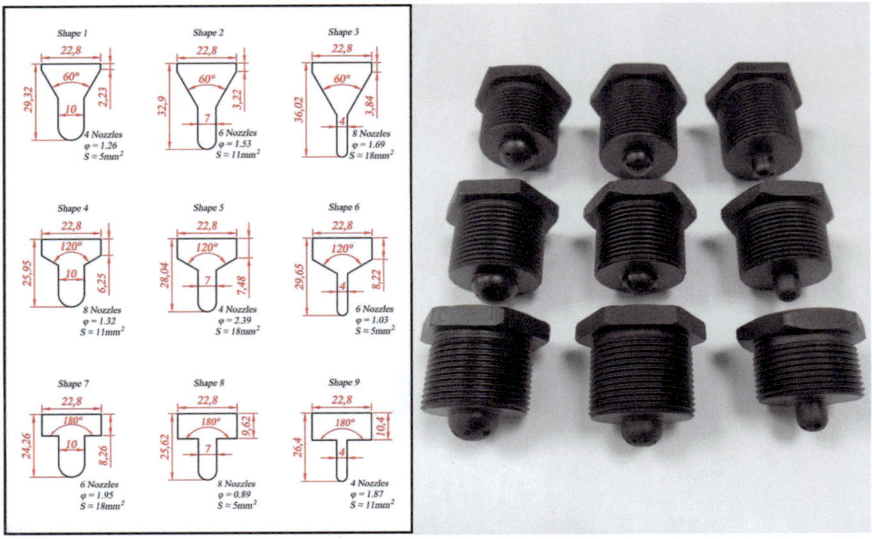

Fig. 3.14 Nine PC redesigned by Taguchi method (Wang et al. 2022)

area, with a weightage of 90%. The lean burn limit was extended for an increase in the total orifice area.

3.5.3 *Engine Studies*

The effect of ON and orientations on the combustion characteristics of a port fuel injection (PFI) SI engine was investigated by Baumgartner et al. (2017). It was found that the total orifice area affects the processes in the PC. The jet ejection was quick for the small orifice area. The authors highlighted the role of jet and wall impingement on the combustion of the charge in the MC. The jet quenches at the wall if the jet travels along the surface and hits the cylinder walls. The impingement of the jet perpendicularly to the surface, followed by the re-entrainment of the jet, showed the enhancement of the combustion for a small orifice size. Guiding the flow of the jet towards the fire deck results in improved spatial distribution. For large cross-sectional areas of the orifice, the jet-surface contact should be avoided by setting the appropriate orifice angle (Baumgartner et al. 2017). Wang et al. (2023) optimised the orifice structure, injection strategy, and piston geometry. The latter two parameters are not in the scope of this chapter. It was found that the orifice structure should be compatible with MC. Optimum OA should allow jets to travel far in the MC without hitting the chamber walls so that the ultra-lean limit is maximized.

Wang et al. (2024) studied the effect of passive PC design parameter effect on the combustion, performance, and emission characteristics of a large-bore direct injection SI engine. PC volume and OD were the parameters selected for the investigation. Figure 3.15 shows the effect of PC volume and OD on in-cylinder pressure and temperature, HRR, and mass fraction burned. Three PCs were named PC1, PC2, and PC3. It was found that the increase in the PC volume deteriorates the in-cylinder combustion and performance of the engine. The HRR was higher and advanced for small PC volume and OD, and the 10–90% mass fraction burned duration was shorter. Large OD improves the combustion stability at high loads but deteriorates at low loads. All the PC designs performed better than spark ignition in terms of ITE, which was more promising at high load. The PC design with a small OD and a small PC volume improves the combustion process and thermal efficiency for the complete engine operation range of load. Antolini et al. (2023) studied the effect of OD on the combustion process for a passive PC fitted in an optical SI engine. Among the three ODs, the smallest one showed a longer duration between spark ignition in PC and jet ejection. However, the combustion in MC for the smallest OD was the shortest due to the high velocity and turbulence generated in the MC.

Hua et al. (2021) investigated the effect of PC volume and number of holes on the performance and emission of SI engines with direct injection in MC. The indicated mean effective pressure (IMEP) and performance of the small PC volume were superior due to low heat losses in the PC. The hot jet from a single-orifice PC showed

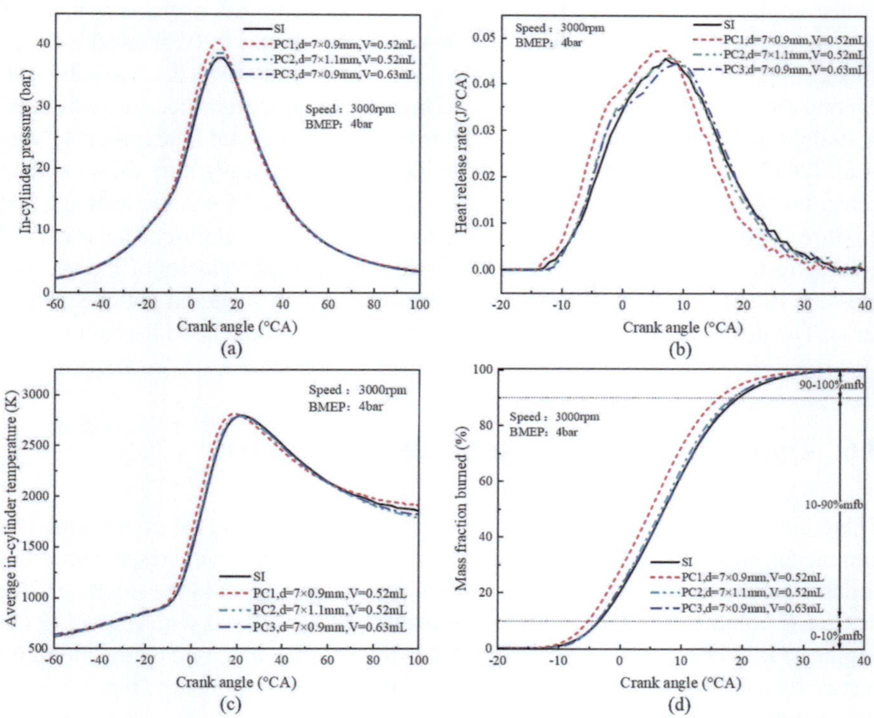

Fig. 3.15 Combustion characteristics for different passive PC designs (Wang et al. 2024)

a better combustion rate than the 7-orifice PC. The HC, OC, and NOx emissions were lower for a small volume PC. The study suggested a single-hole PC for low CO emissions. A lower volume PC and single-hole PC are suitable for lower HC emissions. For the optimized PC, a 92% reduction in NOx emissions was achieved. Zhou et al. (2022) investigated the effect of different orifice sizes of the PC on the performance and emission of an engine. The performance and emission results from different combinations of the designs are shown in Fig. 3.16. For a range of sizes, the engine's performance was better for 3- and 4-mm orifice-size PCs. The large orifice, which is 6 mm in this case, results in low jet velocity. It results in an increase in engine instabilities; hence, fuel consumption was higher.

The small orifice (2 mm) results in high jet velocity, but the throttling loss and quenching make it difficult to ignite the charge in the MC. The NOx emission didn't vary with orifice size for lean engine operation. The HC and CO emissions were lower for 4 and 5 mm orifice sizes. Thus, these sizes proved to have better combustion efficiencies. Novella et al. (2023) optimized the PC geometry for low-speed and high-load engine operating conditions. In the experimental investigation, it was found that the performance of the PC and SI engines was the same for low speed and high load with high knock propensity, with the PC engine due to a drop in TKE and flow velocity. Using 1-D modelling, the authors proposed a comprised PC geometry for the complete engine operating map. Huang et al. (2024) investigated the effect of orifice angle, orifice size, and orifice hole angle on combustion characteristics of PC engines with DI in MC. Among 140 and 120° orifices, 120° showed a better knock suppression limit, rapid combustion, and a lean limit of $\lambda = 3.1$ was achieved. Among the two orifice areas, the combustion rate for the small orifice area was high. A straight orifice hole showed better performance than a tapered hole orifice. Hlaing et al. (2022) optimized PC volume and orifice area for a heavy-duty diesel engine converted to a PC combustion engine. The NOx for all the PCs in the lean air–fuel mixture declined drastically and was dependent on jet temperature. The small PC volume results in higher ITE due to low heat loss and high velocity of the jet. The pressure rise relation with the orifice area was non-linear. It should not be small or large. The details of the PC engine studies can be found in the cited literature.

3.6 Optimum Values of Prechamber Parameters

There are several design parameters of PCs, and some depend on each other. The parameters marked with a star in Table 3.1 are normalized with respect to other suitable parameters. The PC volume is normalised with respect to the engine, RCM, or CVCC volume. The OD and throat diameter are normalised with respect to the engine or RCM bore. The values suggested in this chapter are based on the literature survey to give an idea of the order of magnitude of PC parameters; these are not strictly optimum values, and they may vary for other PC designs.

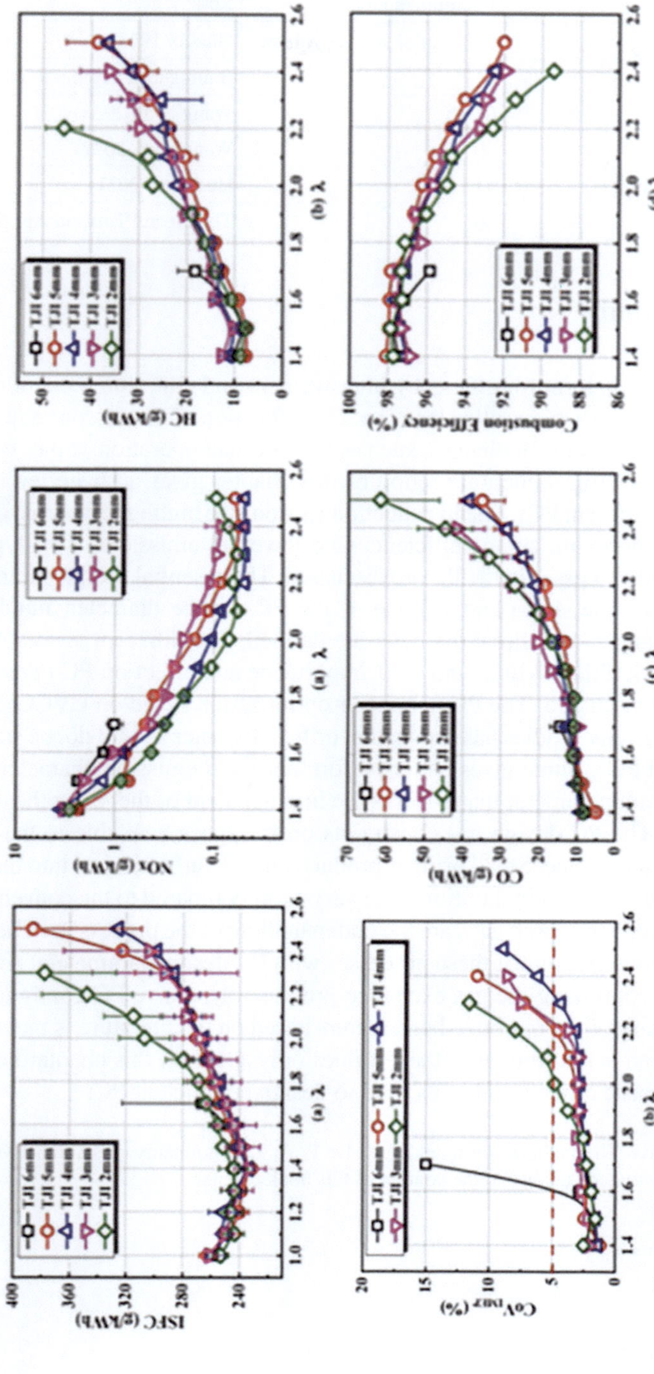

Fig. 3.16 Effect of orifice on performance and emission characteristics (Zhou et al. 2022)

Table 3.1 Optimized prechamber parameters

Parameter	Recommendation	Ref.
PC volume*	2–3% of clearance volume	Gussak 1975)
Orifice number	6	Baumgartner et al. 2017)
Orifice diameter*	0.015	Wang et al. 2023)
Orifice angle	90 °	Wang et al. 2022)
(Orifice area/PC Volume) x Bore	410	Hua et al. 2021)
Transition angle	60°	Thelen and Toulson Apr. 2016)

3.7 Conclusions

The PC concept is helpful for efficiency improvement and emission reduction. The active PC concept can achieve ultra-lean ($\lambda > 2$) combustion of the gaseous and liquid fuels. There are still many challenges, such as the low load operation of passive PCs and scavenging. The introduction section of this chapter gives an overview of the development of different PCs. The introduction section also highlights the role of lean combustion in improving engine efficiency and lowering emissions. Two types of PCs and different processes in the PC are discussed. The essential design parameters of the PC include the shape and volume of the PC, orifice diameter, number of orifices, orifice angle, and throat diameter are defined. The different processes and MC for a PC jet ignition system and their importance and effect on PC parameters in processes are discussed. The PC parameter optimization based on CVCC, RCM, simulation, and engine studies has shown that orifice diameter should not be small or large. The small PC volume gives better performance and emission characteristics. OA is an important parameter that governs the impingement of the jet on the engine cylinder walls. The PC design has constraints on the space available at the spark plug location in the engine. Small orifices produce intense turbulent jets into the MC, which causes the combustion duration to be very short compared to the conventional SI engine. The major research gaps are the understanding of the process in PC and the development of correlations of these processes with PC design parameters. Another research gap is optimizing the gas exchange process for passive PC configuration. Finally, a summary of the optimized parameters based on the literature is mentioned for readers, which is not limited to these values only. Overall, this chapter helps to understand the basic aspects and tools used to design an efficient PC.

Acknowledgements The authors are thankful to Dr P. A. Lakshminarayanan for his valuable suggestions to improve the quality of the content of this book chapter.

References

Adams TG (1979) Torch ignition for combustion control of lean mixtures (No. 790440). SAE Techical Paper

Alvarez CEC, Couto GE, Roso VR, Thiriet AB, Valle RM (2018) A review of prechamber ignition systems as lean combustion technology for SI engines. Appl Therm Eng 128:107–120

Antolini J, Sementa P, Tornatore C, Catapano F, Vaglieco BM, Desantes JM, López JJ (2023) Effect of passive pre-chamber orifice diameter on the methane combustion process in an optically accessible SI engine. Fuel 341:126990

Baumgartner LS, Karmann S, Backes F, Stadler A, Wachtmeister G (2017) Experimental investigation of orifice design effects on a methane fuelled prechamber gas engine for automotive applications (No. 2017–24–0096) SAE Techical Paper

Benajes J, Novella R, Gómez-Soriano J, Martínez-Hernándiz PJ, Libert C, Dabiri M (2019) Evaluation of the passive pre-chamber ignition concept for future high compression ratio turbocharged spark-ignition engines. Appl Energy 248:576–588

Benajes J, Novella R, Gomez-Soriano J, Barbery I, Libert C, Rampanarivo F, Dabiri M (2020) Computational assessment towards understanding the energy conversion and combustion process of lean mixtures in passive pre-chamber ignited engines. Appl Therm Eng 178:115501

Biswas S (2018) Physics of turbulent jet ignition: mechanisms and dynamics of ultra-lean combustion. Springer

Blankmeister M, Alp M, Shimizu E (2018) Passive prechamber spark plug for future gasoline combustion systems with direct injection. In: Ignition systems for gasoline engines: 4th international conference, Berlin, Germany Ed: M G√°nther, p 149

Brandstetter W (1980) The volkswagen lean burn pc-engine concept. SAE Trans 1804–1821

Chinnathambi, P., Bunce, M., and Cruff, L. (2015). RANS based multidimensional modeling of an ultra-lean burn pre-chamber combustion system with auxiliary liquid gasoline injection (No. 2015–01–0386). SAE Technical Paper

Date T, Yagi S, Ishizuya A, Fujii I (1974) Research and development of the Honda CVCC engine (No. 740605). SAE Technical Paper

Distaso E, Amirante R, Cassone E, De Palma P, Sementa P, Tamburrano P, Vaglieco BM (2020) Analysis of the combustion process in a lean-burning turbulent jet ignition engine fueled with methane. Energy Convers Manage 223:113257

Frasci E, Rosa RN, Moreno BP, Arsie I, Jannelli E (2023) Impact of prechamber design and air–fuel ratio on combustion and fuel consumption in a SI engine equipped with a passive TJI. Fuel 345:128265

Gentz G, Gholamisheeri M, Toulson E (2017) A study of a turbulent jet ignition system fueled with iso-octane: pressure trace analysis and combustion visualization. Appl Energy 189:385–394

Guo X, Li T, Chen R, Huang S, Zhou X, Wang N, Li S (2024) Effects of the nozzle design parameters on turbulent jet development of active pre-chamber. Energy 306:132568

Gussak LA (1975) High chemical activity of incomplete combustion products and a method of prechamber torch ignition for avalanche activation of combustion in internal combustion engines. SAE Trans 2421–2445

Gussak LA, Karpov VP, Tikhonov YV (1979) The application of lag-process in prechamber engines. SAE Trans 2355–2380

Gussak LA, Karpov VP, Slutskii VGE, Spasski AI (1983) Burning rate and stability in forechamber flame ignition in an internal-combustion engine. Combust, Explos Shock Waves (Engl Transl);(United States) 19(5)

Heywood JB (2018) Internal combustion engine fundamentals. McGraw-Hill Education

Hlaing P, Marquez ME, Cenker E, Im HG, Johansson B, Turner JW (2022) Effects of volume and nozzle area in narrow-throat spark-ignited pre-chamber combustion engines. Fuel 313:123029

Ho CM, Santavicca DA (1987) Turbulence effects on early flame kernal growth. SAE Trans 505–512

Hu C, Zhang Z, Tian M, Liu N, Wei S (2022) Research on application of asymmetrical Pre-chamber in air-assisted direct injection kerosene engine. Appl Therm Eng 204:117919

Hua J, Zhou L, Gao Q, Feng Z, Wei H (2021) Influence of pre-chamber structure and injection parameters on engine performance and combustion characteristics in a turbulent jet ignition (TJI) engine. Fuel 283:119236

Huang Z, Wang L, Pan H, Chen H, Li J, Wang T, Wang L (2024) Effects of key parameters of pre-chamber on engine combustion and emissions. Therm Sci Eng Prog 49:102450

Idicheria CA, Yun H, Najt PM (2018) An advanced ignition system for high efficiency engines. In: ignition systems for gasoline engines: 4th international conference. Ed.: Günther M, Berlin, Germany vol 6–7, p 40

Jung D, Sasaki K, Iida N (2017) Effects of increased spark discharge energy and enhanced in-cylinder turbulence level on lean limits and cycle-to-cycle variations of combustion for SI engine operation. Appl Energy 205:1467–1477

Koga T, Shuto S, Takemoto D, Oda Y, Yuki A, Furukawa Y (2019) Development of pre-chamber with less combustion variation with use of LES (No. 2019–01–2223) SAE Techical Paper

Krajnović J, Sjerić M, Tomić R, Kozarac D (2024) A novel concept of active pre-chamber engine with a single injector–The passive main chamber approach. Appl Therm Eng 250:123509

Kyrtatos P, Bardis K, Bolla M, Denisov A, Wright Y M, Herrmann K, Boulouchos K (2018) 8.3 transferability of insights from fundamental investigations into practical applications of prechamber combustion systems. In: Ignition Systems for Gasoline Engines: Internationale Tagung Zündsysteme für Ottomotoren, pp 442

Liu Z, Zheng Z (2024) The effect of ignition energy on the lean combustion limitation in high compression ratio engines. Energy 301:131591

Liu X, Marquez ME, Sanal S, Silva M, AlRamadan AS, Cenker E, Im HG (2023) Computational assessment of the effects of pre-chamber and piston geometries on the combustion characteristics of an optical pre-chamber engine. Fuel 341:127659

Lumsden, G., and Watson, H. C. (1995). Optimum Control of an SI Engine with a $\lambda = 5$ Capability (No. 950689). SAE Technical Paper

Lv Y, Feng S, Luo J, Liu Q, Li L, Kang Z (2024) Effect of key structure parameters of passive pre-chamber on in-cylinder combustion processes and emissions within a gasoline engine. Case Stud Therm Eng 59:104467

Müller C, Pischinger S, Tews S, Müller A, Habermann K (2021) Analysis of experimental results with an active pre-chamber ultra-lean burn SI engine. Int J Engine Res 22(10):3103–3127

Nikolai NS (1990) Biographical memoirs of fellows of the royal society, vol 36, pp 526–546

Noguchi M, Sanda S, Nakamura N (1976) Development of Toyota lean burn engine. SAE Trans 2358–2373

Novella R, Gómez-Soriano J, Barbery I, Libert C (2022) Numerical analysis of the passive pre-chamber ignition concept for light duty applications. Appl Therm Eng 213:118610

Novella R, De la Morena J, Pagano V, Pitarch R (2023) Optical evaluation of orifice orientation and number effects on active pre-chamber spark ignition combustion. Fuel 338:127265.Z

Quader AA (1974) Lean combustion and the misfire limit in spark ignition engines. SAE Trans 3274–3296

Serrano D, Zaccardi JM, Müller C, Libert C, Habermann K (2019) Ultra-lean pre-chamber gasoline engine for future hybrid powertrains. In: 14th international conference on engines and vehicles

Silva M, Liu X, Hlaing P, Sanal S, Cenker E, Chang J, Im HG (2022) Computational assessment of effects of throat diameter on combustion and turbulence characteristics in a pre-chamber engine. Appl Therm Eng 212:118595

Silva M, Mohan B, Badra J, Zhang A, Hlaing P, Cenker E, Im HG (2023) DoE-ML guided optimization of an active pre-chamber geometry using CFD. Int J Engine Res 24(7):2936–2948

Srivastava D, Agarwal AK (2013) Laser ignition of single cylinder engine and effects of ignition location. SAE Technical Papers, vol 2. https://doi.org/10.4271/2013-01-1631

Tang Q, Sampath R, Marquez ME, Sharma P, Hlaing P, Houidi MB, Johansson B (2021) Optical diagnostics on the pre-chamber jet and main chamber ignition in the active pre-chamber combustion (PCC). Combust Flame 228:218–235

Thelen BC, Toulson E (2016) A computational study of the effects of spark location on the performance of a turbulent jet ignition system (No. 2016–01–0608). SAE Technical Paper

Thelen BC, Toulson E (2016) A computational study of the effects of spark location on the performance of a turbulent jet ignition system. SAE Technical Papers. https://doi.org/10.4271/2016-01-0608

Toulson E, Huisjen A, Chen X, Squibb C, Zhu G, Schock H, Attard WP (2012) Visualization of propane and natural gas spark ignition and turbulent jet ignition combustion. SAE Int J Engines 5(4):1821–1835

Toulson E, Schock HJ, Attard WP (2010) A review of pre-chamber-initiated jet ignition combustion systems

Turkish MC (1974) 3-valve stratified charge engines: Evolvement, analysis, and progression. SAE Trans 3483–3503

Vedula RT, Song R, Stuecken T, Zhu GG, Schock H (2017) Thermal efficiency of a dual-mode turbulent jet ignition engine under lean and near-stoichiometric operation. Int J Engine Res 18(10):1055–1066

Wang B, Xie F, Hong W, Du J, Chen H, Su Y (2022) The effect of structural parameters of pre-chamber with turbulent jet ignition system on combustion characteristics of methanol-air premixture. Energy Convers Manage 274:116473

Wang B, Xie F, Hong W, Du J, Chen H, Li X (2023) Extending ultra-lean burn performance of high compression ratio pre-chamber jet ignition engines based on injection strategy and optimized structure. Energy 282:128433

Wang R, Zhou F, Liu J, Zhang Q, Qiao J, Jia D, Shen D (2024) Experimental investigation of the influence of pre-chamber structural parameters on combustion and emission characteristics of a high compression ratio SI engine. Therm Sci Eng Prog 47:102318

Zhao D, An Y, Pei Y, Shi H, Wang K (2023) Numerical study on the asymmetrical jets formation from active pre-chamber under super-lean combustion conditions. Energy 262:125446

Zhao D, An Y, Pei Y, Hu J, Hu Z, Zhang Y (2024) Optical study on spray mixing, flame propagation and jets evolution within visible methanol active pre-chamber for turbulent jet ignition. Energy Convers Manag 319:118952

Zhou L, Song Y, Hua J, Liu F, Liu Z, Wei H (2022) Effects of different hole structures of pre-chamber with turbulent jet ignition on the flame propagation and lean combustion performance of a single-cylinder engine. Fuel 308:121902

Zhou L, Ding Y, Li AQ, Song Y, Liu Z, Wei H (2023) Experimental study of gasoline engine with EGR dilution based on reactivity controlled turbulent jet ignition (RCTJI). Fuel 331:125744

Zhu S, Akehurst S, Lewis A, Yuan H (2022) A review of the pre-chamber ignition system applied on future low-carbon spark ignition engines. Renew Sustain Energy Rev 154:111872

Chapter 4
Fuel Properties and Suitability for Internal Combustion Engines: Exploring Conventional, Alternative, and Bio-blended Fuels

Pushpendra Singh, Debarshi Baruah, Siddhartha Kumar Gohain, and Pankaj Kalita

Abstract This chapter delves into the characteristics of various liquid and gaseous fuels, focusing on their properties and their suitability for different internal combustion (IC) engines. Liquid fuels such as gasoline, diesel, kerosene, and ethanol are examined in terms of energy content, volatility, combustion characteristics, and storage requirements. Gaseous fuels, including natural gas, propane, and hydrogen, are analyzed for their calorific value, combustibility, and ease of handling. Essential properties required in an ideal fuel for IC engines include high energy density, appropriate volatility, minimal residue production, and favorable combustion efficiency, ensuring optimal engine performance, reduced emissions, and operational reliability. The chapter outlines the specific requirements of different IC engines, such as spark-ignition (SI) and compression-ignition (CI) engines, where SI engines benefit from fuels with high octane ratings to prevent knocking, and CI engines require fuels with high cetane numbers for efficient ignition. Alternative fossil fuels, such as liquefied petroleum gas (LPG) and compressed natural gas (CNG), are also discussed for their role in reducing dependency on traditional fossil fuels. The chapter further explores bio-blends, including biodiesel, bioethanol, and biogas, highlighting their properties like renewability, lower greenhouse gas emissions, and compatibility with existing engine technologies. Biodiesel's higher lubricity and lower sulfur content make it favorable for CI engines, while bioethanol's high octane rating suits SI engines. This chapter underscores the potential of these bio-blends to enhance energy security and environmental sustainability, demonstrating their growing importance in the future of IC engine fuel technology.

P. Singh · D. Baruah · P. Kalita (✉)
School of Energy Science and Engineering, Indian Institute of Technology Guwahati, Guwahati, Assam, India
e-mail: pankajk@iitg.ac.in

S. K. Gohain
Department of Chemical Engineering, Bineswar Brahma Engineering College, Kokrajhar, Assam, India

© The Author(s), under exclusive license to Springer Nature Singapore Pte Ltd. 2026
H. Valera and A. K. Agarwal (eds.), *Ammonia and Low Carbon Combustion Engine Technologies for De-Fossilizing Transport Sector*, Energy, Environment, and Sustainability, https://doi.org/10.1007/978-981-96-6628-7_4

Keywords Liquid fuels · Gaseous fuels · IC engine · Biofuels · Alternative fuels

4.1 Introduction

The evolution of internal combustion (IC) engine technology has greatly influenced modern transportation, energy production, and various industrial applications. Central to this technological progress is the variety and quality of fuels used in these engines. The properties of liquid and gaseous fuels are crucial in determining engine performance, efficiency, and environmental impact. Traditional fuels like gasoline and diesel have long been the mainstay of IC engines due to their high energy density and widespread availability. However, growing concerns about environmental pollution and the depletion of fossil fuel reserves have spurred interest in alternative fuels, including bio-blends and sustainable fossil fuel alternatives.

Liquid fuels such as gasoline and diesel are valued for their high energy content and relatively simple storage and handling requirements (Das et al. 2022; Bae and Kim 2017). Gasoline, with its high volatility and octane rating, is primarily used in spark-ignition (SI) engines to prevent knocking and ensure smooth combustion (Demirbas et al. 2015; Ugurlu and Oztuna 2015). Diesel, on the other hand, is preferred for compression-ignition (CI) engines due to its higher energy density and cetane number, which promote efficient ignition and combustion under high-pressure conditions (Pham et al. 2022). Other liquid fuels like kerosene and ethanol offer various benefits and are used in specific applications based on their unique properties (Pham et al. 2022; Astbury 2008). The rise in energy consumption, generated by different fossil fuels, is shown in Fig. 4.1.

Global energy consumption for all transportation modes is expected to grow from 2012 to 2040, as shown in Fig. 4.2. Passenger travel energy use increases by 1.4% annually, rising from 63 quadrillion Btu in 2012 to 94 quadrillion Btu in 2040. Freight-related energy consumption grows by 1.5% annually, reaching 60 quadrillion Btu by 2040.

Gaseous fuels, including natural gas, propane, and hydrogen, present different characteristics. These fuels are known for their high combustibility and clean-burning properties, making them attractive options for reducing emissions from IC engines (Ilhak et al. 2019; Algayyim et al. 2024). Natural gas, for instance, has a high methane content, allowing for efficient combustion with lower carbon dioxide emissions compared to gasoline and diesel (Civan 1989). Propane and hydrogen offer similar benefits, with hydrogen being particularly notable for its potential as a zero-emission fuel when produced from renewable sources (Bae and Kim 2017).

Various research is ongoing on alternative liquid and gaseous fuels to reduce the dependency on fossil-based fuels. Higher linear and aromatic alcohols were reported by Hua (Hua 2024) as alternative fuels for IC engines. It was reported that higher alcohols enhance biodiesel performance and emissions more effectively than diesel, with RCCI engines benefiting from reduced PM and NOx emissions. Sedghi et al. (2022) explored converting glycerol into fuels or additives to enhance

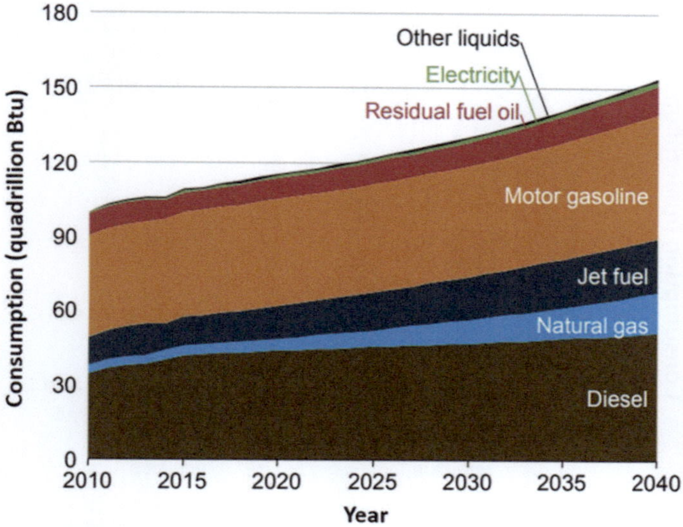

Fig. 4.1 Rise in global consumption of different fuels in the transportation sector (EIA. International Energy Outlook 2016)

IC engine performance. Various conversion routes and their effects on engines were discussed, emphasizing that direct glycerol use is problematic, but glycerol derivatives show promise for future research. Paykani et al. (2022) examined syngas's impact on performance and emissions in various engine types and explored on-board fuel reforming techniques. The review highlights the importance of production methods, base fuel choice, and engine conditions in optimizing syngas use for cleaner combustion and increased efficiency. Awad et al. (2018) reviewed the use of alcohols (butanol, methanol, ethanol, and fusel oil) and ethers as fuels in spark ignition engines, driven by energy security and global warming concerns. The study highlights that these fuels improve combustion efficiency and reduce emissions due to their higher oxygen content, though they result in lower energy values and increased fuel consumption compared to gasoline. Some of the existing work on the use of different fuels in IC engines for reducing the emission of greenhouse gases is shown in Table 4.1.

The need for sustainable and environmentally friendly fuels has led to the development and adoption of bio-blends, such as biodiesel and bioethanol. Biodiesel, derived from vegetable oils or animal fats, provides higher lubricity and lower sulfur content, making it suitable for CI engines (Sivaramakrishnan and Ravikumar 2012; Verma et al. xxxx; Sahu and Shukla 2023). Patil et al. (2024) focus on engine deposits, wear, and lubricating oil degradation, noting that biofuel blends can increase deposits and wear but also offer benefits due to their oxygenated nature. Studies show mixed results, with biodiesel blends reducing deposits in some cases while increasing corrosion and lubrication oil degradation. Bioethanol, with its high octane rating, is increasingly used in SI engines either as a pure fuel or in blends with gasoline to

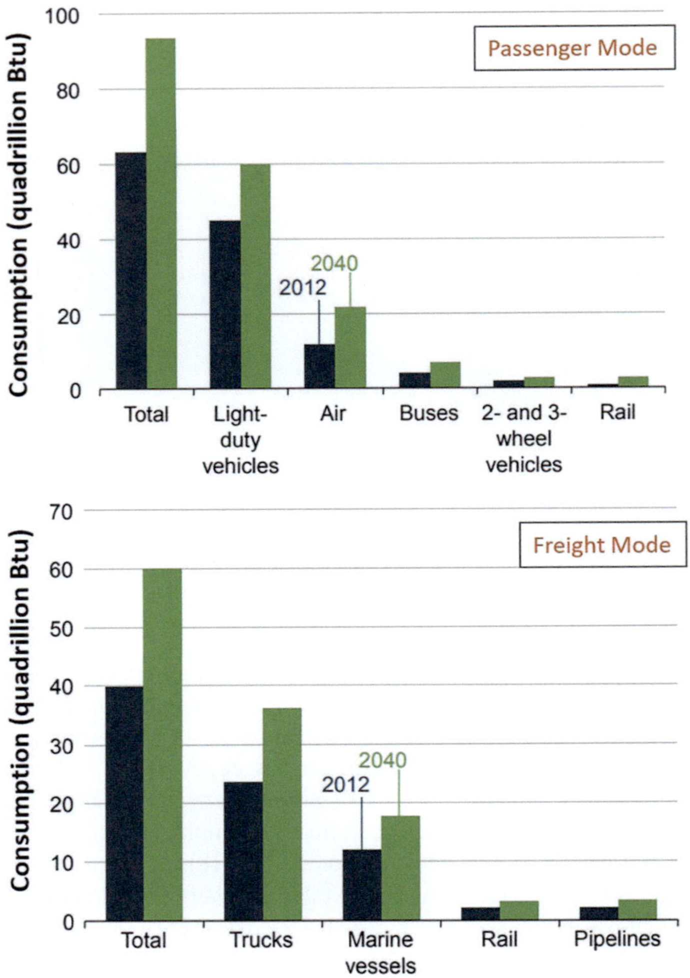

Fig. 4.2 Expected rise in fuel consumption by different modes of transportation (EIA. International Energy Outlook 2016)

enhance performance and reduce greenhouse gas emissions (Jain and Kumar 2024; Oktem and Oktem 2024).

This chapter provides a detailed overview of the characteristics and properties of various liquid and gaseous fuels, emphasizing the essential attributes required for optimal performance in IC engines. It highlights the role of alternative fossil fuels and bio-blends in advancing contemporary IC engine technology. By exploring these factors, the chapter aims to offer a deeper understanding of advancements in fuel technology and their implications for future engine development and environmental sustainability. The analysis focuses on key aspects such as energy content, combustion efficiency, emissions, and compatibility with existing engine designs,

Table 4.1 Previous research on the application of different fuels in IC engine

Author (Year)	Fuel used	Type of engine	Benefit of using this fuel in an IC engine
Wu et al. (2024)	Ammonia-gasoline	Four-cylinder, naturally aspirated, SI engine	Carbon reduction potential with up to 46.8% CO_2 reduction at 35% AER; maintains engine power with minimal structural changes
Yakin et al. (2024)	Ammonia-borane blends	Not specified	Reduced NOx (up to 23%), CO (up to 88.2%), and HC emissions; potential for lower exhaust temperatures
Li et al. (2024a)	Polyoxymethylene dimethyl ethers (PODE), gasoline, diesel	Diesel engine compliant with China VI standards	Enhanced combustion efficiency, reduced CO, THC, and PN emissions; balanced reduction in NOx emissions with 5% gasoline and 20% PODE blend
Li et al. (2024b)	Diesel-methanol, isopropanol-butanol-methanol (IBE) blends	Diesel-methanol dual-fuel engine	Significant soot emission reduction; improved CO, HC, and NOx emissions with $\leq$20% IBE in diesel
Ma et al. (2024)	Ammonia-diesel, natural gas	Ammonia-diesel dual-fuel engine	Increased thermal efficiency (up to 51.0%), reduced N_2O by 97%, GHG by 57.2%, and nearly eliminated ammonia slip
Ge et al. (2024)	Diesel-bioethanol blends	Diesel engine	Reduced NOx emissions by over 52% and smoke by over 78%; improved emissions under idle conditions
Zareei et al. (2024)	Hydrogen-enriched compressed natural gas (HCNG), CNG	Modified four-cylinder engine with direct injection	Reduced CO/HC emissions (up to 14%); increased NOx due to higher exhaust temperatures
Zhou et al. (2024a)	Natural gas, diesel with additives (argon, hydrogen, syngas)	Natural gas/diesel dual-fuel RCCI engine	Increased gross indicated efficiency (38.1% to >50%) with reduced low-load HC emissions; argon has the greatest potential for performance improvement

(continued)

Table 4.1 (continued)

Author (Year)	Fuel used	Type of engine	Benefit of using this fuel in an IC engine
Nyongesa et al. (2014)	Natural gas-diesel dual-fuel	Low-pressure gas-admission marine two-stroke engine	Improved combustion efficiency and emissions under low-load conditions; reduced unburned CH_4 (20.14%) and CO emissions (42.29%)
Gültekin (2024)	Hydrogen, diesel, bioethanol (triple fuel mixture)	Hydrogen-diesel dual-fuel compression ignition engine	Reduced specific fuel consumption (5.2%), soot emissions (69%), and HC emissions (29.7%) with a 10% increase in valve lift at 9 Nm load
Shirneshan et al. (2025)	Diesel, biodiesel, n-butanol, ethanol, hydrogen	Single-cylinder diesel engine	Enhanced power output and thermal efficiency with increased hydrogen proportion; alcohol blends reduce CO_2 emissions and NOx, but affect engine performance
Gad and Ismail (2021)	Waste cooking oil (WCO) biodiesel, gasoline, kerosene	Diesel engine	Improved CO (reduced by up to 40%), HC (reduced by up to 43%) emissions, and smoke emissions (reduced by up to 44%) with slight decreases in thermal efficiency
Abuelnour et al. (2024)	Hydrogen, exhaust gas recirculation (EGR)	Diesel-Hydrogen dual-fuel engine	Improved engine efficiency with CO and soot emissions reduced by up to 33% and 88%, respectively
Bhowmi et al. (2024)	Diesel, kerosene, ethanol, hydrogen	Compression ignition engine (modified for hydrogen injection)	Improved brake thermal efficiency (up to 3.2%), brake specific energy consumption (up to 3.1%), lower HC (up to 50.1%), CO (up to 50.5%), and particulate matter (up to 44.8%) emissions
Singh et al. (2024)	Hydrogen-enriched bio-CNG (HBCNG)	Single-cylinder, four-stroke CRDI diesel engine	At high load, a 40% H_2 + 60% BCNG mixture increases

thereby offering valuable insights into how these fuels can address the demands of modern and future IC engines.

4.2 Properties of Liquid and Gaseous Fuel

4.2.1 Gross Calorific Value (GCV)

GCV is defined as the energy content of a fuel, which is determined by the heat generated by a unit fuel sample during combustion in an oxygen-rich environment at a pressure of around 20–40 atm (Kiang 2018). Combustion typically produces water vapour, and the difference between Gross Calorific Value (GCV) and Net Calorific Value (NCV) lies in how the heat of the water vapour is treated. GCV, also known as the High Heating Value (HHV), includes the heat released during the condensation of water vapour (Paulo et al. 2022), whereas NCV does not account for the heat of the steam, as it is allowed to escape. The relationship between HHV and LHV can be defined as follows (Kiang 2018):

$$HHV = LHV + H_v \times \frac{Mass\ of\ water\ vapour\ in\ the\ product\ gas}{Total\ fuel\ mass} \tag{4.1}$$

Here, H_v is the heat of vaporization of water.

A higher calorific value fuel can produce more heat during combustion, thus generating more energy, which ultimately either increases the power output or decreases the fuel consumption. Thus, a high calorific value fuel is preferred to increase the efficiency of the engine. Fuels with higher calorific values are found to be emitting less emission, since they burn very efficiently.

Cloud Point and Pour Point

The cloud point is defined as the temperature at which the first crystals or wax appear in a fuel; it is also known as the Wax Appearance Temperature (WAT) (Eyankware et al. 2016). This occurs due to the solidification of paraffinic components in the fuel mixture, leading to issues such as thickening of the fuel and clogging of components like fuel filters and injectors. Therefore, a higher cloud point indicates greater susceptibility to wax formation. The cloud point is typically measured using the ASTM D2500 Cloud Point Test Apparatus. The pour point is the minimum temperature at which the fuel ceases to flow. Below this temperature, the fuel oil cannot be pumped or moved. Thus, both the cloud point and pour point are important parameters for assessing a fuel's performance in cold weather. To lower the pour point, depressants made from polymethacrylates are used (Shah and Schevon 2022).

Octane/Cetane Number

The octane number, or octane rating, is a key measure of gasoline or fuel used in high-speed engines. It indicates the fuel's resistance to knocking or premature ignition. The octane number represents the percentage by volume of iso-octane in a mixture of iso-octane and heptane that matches the performance of the tested fuel in a standard test engine. Tetraethyl lead (TEL) was once commonly used as an additive to improve the octane rating, but it was replaced by Methyl Tertiary-Butyl Ether (MTBE) because TEL generated lead oxides during combustion, which poisoned catalytic converters (Demirbas et al. 2015).

The cetane number of a fuel is defined as the percentage by volume of n-cetane in a mixture of n-cetane and α-methyl naphthalene that matches the ignition characteristics of the fuel in a standard test engine operating under specified conditions (Sivaramakrishnan and Ravikumar 2012). The cetane number measures ignition delay; a lower cetane number indicates a longer delay in combustion, which can lead to incomplete combustion, smoke, and poor fuel ignition. A fuel with a high cetane number has good ignition quality, meaning the ignition delay period between the start of fuel injection and the onset of auto-ignition is short.

Density (P)/API Gravity

The density of fluid generally represents the quantity of fluid contained in a unit volume. Liquid fuels are incompressible in nature, and their density is determined as

$$\rho_l = \frac{m_f}{V_f} \tag{4.2}$$

Here, m_f and V_f are the mass (kg) and volume (m^3) of fluid, respectively.

The gases are compressible in nature, so considering the compressibility factor, the density of gaseous fuel can be determined as

$$\rho_g = \frac{MP}{ZRT} \tag{4.3}$$

Here, M is the molecular weight of gas in kg/mol, P is the absolute pressure in bar, R is the gas constant, T is the absolute temperature in K, Z is the compressibility factor.

American Petroleum Institute (API) gravity is used to describe how dense or viscous a fuel is with respect to water at 60 °F. API gravity is calculated as

$$\text{API gravity at } 60\,^\circ\text{F} = \frac{141.5}{\text{specific gravity at } 60\,^\circ\text{F}} - 131.5 \tag{4.4}$$

Based on the density and API gravity value, the crude oil is classified into the following categories, mentioned in Table 4.2 (Correa Pabón et al. 2019).

Table 4.2 Classification of crude oil

Classification	API scale	Density (kg/m^3)
Light	>31.1	<870
Medium	22.3–31.1	920–870
Heavy	10–22.3	920–1000
Extra heavy	<10	>1000

Specific Gravity

Specific gravity (S) is a dimensionless number expressing the density of a given fluid $\left(\rho_{fluid}\right)$ relative to the density of a reference fluid $\left(\rho_{ref.}\right)$ at the same pressure and temperature conditions (Civan 1989). In the case of gaseous fluid, usually, air at 1 atm and 15.56 °C (60 °F) is used as a reference fluid (Dembicki 2017). In the case of liquid fuel, water at 4 °C (39.2 °F) is considered a standard fluid. Mathematically, it is given by

$$S = \frac{\rho_{fluid}}{\rho_{ref.}} \tag{4.5}$$

Flash Point and Fire Point

The flash point is the minimum temperature at which a fuel produces enough vapor to form a vapor-air mixture that can ignite when exposed to a source of ignition, though the vapor may not be sufficient to sustain continuous burning. The fire point is the temperature at which the fuel produces enough vapor to burn continuously for at least five seconds when exposed to an ignition source (Ljubas et al. 2010). These temperatures define the ignition characteristics of fuel and are crucial safety considerations during storage, transportation, and use. There are three primary methods for determining the flash point: the Closed Cup Abel–Pensky method, the Closed Cup Pensky–Martens method, and the Open Cup Cleveland method.

Sulfur Content

Sulfur is naturally present in crude oil, with low levels referred to as sweet crude and moderate to high levels as sour crude. During the refining process, this sulfur can carry over into various streams and, if not treated, can also be found in finished products like gasoline and diesel. When sulfur is burned, it produces oxides, which are pollutants known to cause acid rain. Sulfur can also lead to corrosion in engines. Currently, the sulfur content in fuel is limited to less than 10 ppmw, often referred to as "zero sulfur" (Ito and Veen 2006).

Viscosity

Viscosity is the measure of a fuel's resistance to flow. It is a critical factor as it influences the movement and injection of fuel into the engine. Typically, kinematic viscosity is used to describe the viscosity of a fuel or oil (Oasmaa et al. 1997). The relationship between kinematic and dynamic viscosity is given by

$$kinematic\ viscosity = \frac{dynamic\ viscosity}{density} \tag{4.6}$$

Carbon Residue

Carbon residue refers to the tendency of a fuel to form carbon deposits during combustion, particularly under high temperatures in inert conditions. Fuels with a high carbon-to-hydrogen ratio are more challenging to burn completely, leading to increased deposits in combustion and exhaust particles. Carbon residue is generally measured by weight percentage or in parts per million (ppm), with high values being undesirable. Carbon residue can also be determined through thermogravimetric (TG) analysis (Ghetti 1994).

Flammability Limits

Flammability limits refer to the range of gas concentration in the air over which a fuel-gas mixture is flammable. This is an important safety consideration because when liquid or gaseous fuel is spilled, its concentration at the spill site is initially 100%. Over time, the fuel diffuses and spreads throughout the surrounding area, creating a concentration profile. At some point, the fuel becomes so diluted that it falls below the flammable limit (Astbury 2008). The combustion hazard of fuel-air mixtures increases with a higher volume fraction of added fuel gases but decreases with higher relative humidity in the air (Wang et al. 2020).

Auto Ignition Temperature

The auto-ignition temperature is the temperature at which a gaseous or vapor mixture ignites spontaneously in a confined space. For any specific mixture of fuel, oxidizer, and diluent, temperature and pressure are the key parameters driving auto-ignition (Askari et al. 2017). According to J. Colson et al. humidity does not have a significant effect on the auto-ignition of gases, although the magnitude of the effect was found to be limited (Johnson and Mashuga 2023).

Boiling Point

The boiling point is a crucial factor in the separation of fuel fractions from crude oil. A large difference in boiling points between the solvent and the mixture is desirable for efficient product separation and solvent regeneration through distillation and stripping (Reddy and Letcher 2007). Since boiling point is directly related to a fuel's volatility—meaning that fuels with lower boiling points evaporate more easily—spilled fuel with a low boiling point will diffuse quickly, reducing the persistence of the spill (Astbury 2008).

Flame Temperature

Flame temperature is a fundamental characteristic of fuel, providing insight into the quality of combustion. A higher flame temperature generally indicates a complete combustion process, which is desirable as it yields greater useful energy. However, it is also observed that lower flame temperatures can be effective in reducing pollutant formation, especially nitrogen oxides (NOx) (Lee et al. 2021).

4.3 Fuel Requirements for Different IC Engines

Different types of internal combustion (IC) engines have specific fuel requirements to achieve optimal performance and efficiency. This section discusses the distinct fuel properties necessary for spark-ignition (SI) engines and compression-ignition (CI) engines.

Spark-Ignition (SI) Engines

SI engines primarily operate on gasoline, which requires a high octane rating to prevent knocking and ensure smooth combustion. The octane number measures a fuel's ability to resist pre-ignition, which is critical for maintaining engine performance and preventing damage (Demirbas et al. 2015). High-volatility fuels are preferred for SI engines to ensure rapid vaporization and efficient air-fuel mixing. Gasoline's properties, including its high energy density and ability to provide quick starts and responsive acceleration, make it ideal for SI engines used in passenger vehicles and light-duty trucks.

A good SI engine fuel must possess the following properties: (Pourkhesalian et al. 2010)

- High octane rating to resist knocking and ensure smooth combustion.
- High volatility for rapid vaporization and efficient air-fuel mixing.
- High energy density to provide sufficient power and fuel efficiency.
- Clean combustion with minimal deposits and pollutants to maintain engine longevity and environmental compliance.
- Consistency and stability for reliable performance under various operating conditions.
- Ease of storage and handling for safe and convenient everyday use.

Gasoline is the most suitable fuel for SI engines due to its high octane rating, volatility, and energy density, which align well with the operating requirements of these engines. Ethanol, often used as a blend with gasoline, is also compatible due to its high octane number, though it has lower energy density (Caton 2010). On the other hand, diesel fuel is unsuitable for SI engines because of its low volatility and high cetane number, which make it more suitable for compression-ignition engines. Diesel's properties would lead to poor vaporization and air-fuel mixing in an SI engine, causing inefficient combustion and potentially damaging the engine. Similarly, fuels like kerosene and heavy oils, which have low volatility and are designed for different types of combustion systems, are not suitable for SI engines.

Compression-Ignition (CI) Engines

CI engines, also known as diesel engines, rely on fuels with a high cetane number, which ensures shorter ignition delays and efficient combustion under high-pressure conditions. Diesel fuel's higher energy density compared to gasoline results in better fuel economy and power output, making it ideal for heavy-duty applications such as trucks, buses, and industrial machinery. Additionally, diesel's lower volatility reduces the risk of vapor lock and enhances safety during storage and handling.

The following properties are desirable in good CI engine fuels (Ilhak et al. 2019; Hua 2024):

- High cetane number to ensure quick ignition and efficient combustion.
- High energy density for better fuel economy and greater power output.
- Low volatility to minimize the risk of vapor lock and improve safety.
- Clean combustion to reduce soot and particulate emissions.
- Stability under varying temperature and pressure conditions.
- Reliable performance in demanding and heavy-duty applications.

Diesel is the most suitable fuel for CI engines due to its high cetane number, which is crucial for the compression-ignition process. Its high energy density also supports the powerful performance required for heavy-duty vehicles and machinery. Other fuels like biodiesel can also be used in CI engines, as they generally have similar properties to conventional diesel, though their lower energy content may slightly affect fuel economy (Wan Ghazali et al. 2015). However, gasoline is unsuitable for CI engines because of its low cetane number and high volatility, which would lead to poor ignition characteristics, inefficient combustion, and potential engine damage. Similarly, fuels like propane or natural gas, which are more suited to spark-ignition engines due to their combustion properties, are not ideal for CI engines.

4.4 Conventional Liquid and Gaseous Fuels

Liquid fuels are fundamental to the IC engine industry due to their high energy density, ease of storage, and efficient transportation. Meanwhile, gaseous fuels are gaining increasing attention in IC engine technology for their high efficiency and cleaner combustion properties. The properties of conventional liquid fuels are shown in Table 4.3, and those of gaseous fuels are shown in Table 4.4. The characteristics of various liquid fuels, including gasoline, diesel, kerosene, and ethanol, as well as gaseous fuels like natural gas, propane, and hydrogen, are discussed below.

Table 4.3 Properties of different conventional liquid fuels

Fuel property	Petrol	Diesel	Bio-diesel	Methanol	Kerosene	Ethanol
GCV (MJ/kg)	46	43	39–41	22.9	37.9	23.4
Cloud point (°C)	–	−3	−3 to 12	–	−21	–
Pour point (°C)	–	−12	−15 to 16	>50	−25	–
Octane rating	85–95	–	–	108.7	15–20	108.6
Cetane rating	17	46–53	47–51	3	42	5–15
API gravity	62.4	30–42	29.29	46.78	41.5	–
Density at 15 °C (g/cm^3)	0.72–0.775	0.84	0.88	0.794	0.79–0.8	0.789
Viscosity at 40 °C (mm^2/s)	6.0	2–4.5	6.0–2.6	–	1.389	–
Flash point (°C)	−43	63	130	11–12	38–72	13
Fire point (°C)	−13	93	100–170	41–42	68<	43
Sulfur content (ppm)	<10	<10	<15	<0.5	<15	–

Table 4.4 Properties of different conventional gaseous fuels

Fuel properties	GCV (kJ/kg)	Specific gravity	Flammability (%)	Auto ignition temperature (°C)	Boiling point (°C)	Flame temperature (°C)
Hydrogen	141.86	0.0696	4–75	500	−253	2045
Methane (Natural Gas)	55.53	0.540	5.3–15	580	−161.5	1957
Butane	49.1	2.01	1.8–8.5	365	−0.5	1970
Propane	50.36	1.52	2.2–9.3	450	−42	1980
Petroleum gas	46–51	0.53–0.54	1.5–9.0	410–580	−27 to 20	1967

Gasoline

Gasoline is the most widely used fuel for spark-ignition (SI) engines, largely due to its high volatility and octane rating, which prevent engine knocking and ensure smooth combustion. The octane rating of gasoline, typically ranging between 87 and 93, measures its ability to resist pre-ignition during combustion, a critical factor in maintaining engine performance and longevity (Awad et al. 2017). Gasoline's high energy density (around 44 MJ/kg) provides the necessary power output for a wide range of vehicles, from small cars to large trucks.

Composed primarily of hydrocarbons, gasoline typically contains a mix of paraffins, olefins, naphthenes, and aromatics, with additives like ethanol and detergents to enhance performance and reduce emissions. The fuel is produced through the refining of crude oil, involving processes like distillation, cracking, and blending to achieve the desired properties. Gasoline's ability to vaporize easily at ambient temperatures makes it highly suitable for cold-start conditions, ensuring reliable engine operation in diverse climates. This combination of characteristics: energy density, volatility, and ease of production, makes gasoline an essential fuel for modern transportation.

Diesel

Diesel fuel is the preferred choice for compression-ignition (CI) engines due to its higher energy density and cetane number, which facilitate efficient ignition under high-pressure conditions (Sivaramakrishnan and Ravikumar 2012). The cetane number of diesel typically ranges from 40 to 55, indicating the fuel's ignition delay; higher cetane numbers signify shorter delays and better ignition quality. Diesel's energy density is about 45–47 MJ/kg, providing superior fuel efficiency and power output, which makes it ideal for heavy-duty applications such as trucks, buses, and industrial machinery.

Diesel fuel is composed primarily of long-chain hydrocarbons and is produced through the refining of crude oil, involving processes like distillation and hydrocracking to break down larger molecules into usable components (Das et al. 2022). Diesel's lower volatility compared to gasoline reduces the risk of vapour lock and makes the fuel safer to handle and store. The fuel's robustness and durability also contribute to its widespread use in vehicles and machinery that demand reliable performance and longevity.

Kerosene

Kerosene is a versatile liquid fuel used extensively in aviation and heating applications. It is characterized by moderate volatility and a high flash point, which enhances safety during handling and storage (Kerosene and Encycl. 2023). Kerosene-based jet fuels, such as Jet-A and Jet-A1, are preferred in aviation due to their stability and energy density, essential for reliable performance at high altitudes. Jet-A and Jet-A1 have a flash point of around 38–72 °C and an energy density of approximately 43–46 MJ/kg.

Kerosene's composition primarily includes hydrocarbons with 10–16 carbon atoms per molecule. It is derived from the distillation of crude oil, where it is separated from other components through fractional distillation. The fuel's ability to burn cleanly and produce a steady heat output makes it suitable not only for aviation but also for residential and industrial heating purposes (Dagaut and Cathonnet 2006). In these applications, kerosene's consistent performance and relatively low cost contribute to its widespread use.

Ethanol

Ethanol is an alcohol-based fuel that is increasingly used either as an additive to or a substitute for gasoline in spark-ignition (SI) engines. With an octane rating of around 100, ethanol enhances engine performance by preventing knocking, similar to gasoline. Its oxygen content promotes more complete combustion, which helps to reduce emissions of carbon monoxide and other pollutants (Lee et al. 2021). Ethanol is commonly blended with gasoline in various proportions, such as E10 (10% ethanol) and E85 (85% ethanol), offering a renewable and environmentally friendly alternative to pure gasoline. However, ethanol has a lower energy density, about 30 MJ/kg, compared to gasoline's 44 MJ/kg, which results in slightly reduced fuel economy and requires more frequent refuelling. The production of ethanol involves fermenting plant materials, such as corn or sugarcane, which are then distilled to separate the ethanol from the fermentation by-products. The various properties associated with different conventional liquid fuels (Wan Ghazali et al. 2015; Inspire SGS. Methanol: Properties 2020; Han and De 2015; Coutinho et al. 2002; Roman et al. 2024) are shown in Table 4.3.

Natural Gas

Natural gas, primarily composed of methane (CH_4), is a popular fuel for internal combustion (IC) engines due to its high energy content and clean combustion profile. Methane's high energy content, approximately 55–57 MJ/kg, allows for efficient combustion, resulting in lower emissions of carbon dioxide, nitrogen oxides, and particulate matter compared to traditional liquid fuels like gasoline and diesel. Natural gas has a high octane rating, around 120–130, which reduces the likelihood of knocking in SI engines, thus enhancing engine performance and longevity (Algayyim et al. 2024; Search 2018). It can be stored and transported in compressed (CNG) or liquefied (LNG) forms, providing flexibility for both stationary and mobile applications. CNG and LNG are produced by processing natural gas extracted from underground deposits, which is then purified and compressed or liquefied to facilitate storage and transport. The clean-burning characteristics of natural gas make it an attractive option for reducing emissions and promoting environmental sustainability in various applications.

Propane

Propane, also known as liquefied petroleum gas (LPG), is a versatile gaseous fuel used widely for various applications. It is favored for its high energy density, approximately 46 MJ/kg, and ease of liquefaction, which makes storage and transportation convenient. Propane burns cleaner than gasoline and diesel, producing fewer greenhouse gases and pollutants such as carbon monoxide and unburned hydrocarbons. This clean-burning characteristic makes propane an attractive option for reducing environmental impact, especially in urban areas where air quality is a significant concern. Additionally, propane has a high octane rating, typically around 104–110, which makes it suitable for SI engines by providing smooth and efficient combustion. Propane is extracted from natural gas processing or crude oil refining and is stored in pressurized tanks as a liquid for ease of handling and use (Pourkhesalian et al. 2010).

Hydrogen

Hydrogen is increasingly recognized as a promising fuel for the future, particularly in the pursuit of sustainable and clean energy solutions. Hydrogen boasts the highest energy content per unit weight, about 120–142 MJ/kg, while its combustion produces only water vapour, making it a zero-emission fuel when derived from renewable sources (Algayyim et al. 2024). Hydrogen can be utilized in both SI and CI engines, although its high diffusivity and low ignition energy present unique challenges in engine design and safety. For SI engines, hydrogen's high octane rating and clean combustion contribute to high power outputs with minimal environmental impact. For CI engines, modifications are often necessary to handle hydrogen's different combustion characteristics. Hydrogen is primarily produced through methods like electrolysis of water or steam methane reforming, and its storage and distribution require advanced technologies due to its low density and tendency to diffuse rapidly (Abubakar et al. 2024). As a fuel, hydrogen represents a key player in the transition to greener internal combustion engine technologies and sustainable energy systems. The properties of various gaseous fuels (Bae and Kim 2017; Algayyim et al. 2024; Rathore et al. 2009) are shown in Table 4.4.

4.5 Alternative Fossil Fuels

Compressed Natural Gas (CNG)

CNG or Compressed Natural Gas is the densified form of natural gas, which is obtained as a fossil fuel. The natural gas obtained directly from gas wells or as a by-product in the exploration of crude oil is compressed up to a pressure of around 200 bar so as to occupy minimum volume at room temperature (Leonardo and Semin 2022). The main constituent is methane (approximately 95%) and the rest may be small amounts of other components like ethane, propane, butane, hexane, CO_2, N_2, etc. (Tahir et al. 2015). For running CNG in SI engines, there are two options: either convert the existing petrol-based SI engine to a dual-fuel engine with retrofit of a CNG kit or utilize specifically designed SI engines to run on CNG. Considering that CNG has a high octane rating of 130, the first option causes unoptimized results. For diesel engines, again there are two options: either run on dual-fuel mode with up to 75% diesel replacement or convert the high compression CI engine to SI with the retrofit of an ignition system (Semin 2008). CNG can be considered a cleaner fuel as compared to conventional gasoline or diesel. A study conducted in Pakistan (Yasar et al. 2013) monitored emissions in different categories of vehicles while converting from diesel or gasoline to CNG. They noted a marked decrease in hydrocarbons, nitrogen oxides and smoke opacity when diesel vehicles were shifted from diesel to CNG. On the other hand, they found that on replacing gasoline with CNG, the hydrocarbons, smoke opacity, sulfur dioxide, and carbon monoxide decrease in the emissions, but there is also a slight increase in the NO part. Kar et al. (2023) stated that approximately 35,00,000 CNG vehicles are plying on Indian roads as of 2023. They also mentioned the possibility of introducing hydrogen with CNG for better results and bolstering the hydrogen economy in India.

Liquefied Natural Gas (LNG)

LNG or Liquefied Natural Gas has emerged as a pivotal fuel in the energy sector, particularly for heavy transportation and maritime applications. LNG is basically natural gas that has been cooled to around −162 °C (−260°F), condensing it into a liquid form that is approximately 600 times smaller in volume than its gaseous state. This compact form makes LNG a practical option for storage and long-distance transport in cryogenic containers (Choi 2011). For engine applications, LNG serves as an alternative to diesel, having several environmental benefits. LNG engines are designed to combust the gas more cleanly than diesel engines, leading to reduced emissions of harmful components and GHGs. According to a study (Livaniou and Papadopoulos 2022), LNG replacement from conventional fuel in ships can result in a drastic reduction of NOx, SOx, and particulate matter, whereas CO_2 reduction in the emissions is not quite significant. The only downside is a slight increase in CO and CH_4 emissions. Arefin et al. (2020), in their review of usage of LNG in dual fuel engines, have stated that LNG is an attractive option from economical, technological and environmental point of view but the challenges are its high flammability posing as a potential hazard and the knocking it produces when introduced in a dual fuel engine. Ghosh et al. (2024) have utilized multiple forecasting techniques using different scenarios to estimate the consumption of LNG in India by 2050, and each of their results predicts a shortfall of 30–50%, which will have to be met by imports. They have emphasized the need for government policies to increase domestic production of LNG to meet this demand by the mentioned time frame.

Liquefied Petroleum Gas (LPG)

LPG or Liquefied Petroleum Gas is again a by-product of oil and gas production. Around 60% of the world's LPG production is by the extraction of crude or natural gas, whereas the remaining 40% is generated from the refining process of crude. The main components are propane, n-butane, isobutane, butene, and propene; the percentage of each component may vary from place to place (Thomson et al. 2003). LPG as a vehicular fuel is not widespread in a country like India, mainly due to the high demand for LPG as an urban cooking fuel. However, LPG as a vehicular fuel is very popular in developed countries. For example, as per 2016 reports, there are approximately 15 million LPG vehicles operating in the European Union (Kontses et al. 2020). Performance-wise, LPG fuel in IC engines does not degrade the power generated substantially. In an experimental study performed on a single-cylinder SI engine with LPG fuel, the results indicated a slight decrease in the power output (approx. 4%) whereas the specific fuel consumption and the fuel pricing difference was noticeably high (approx. 28% and 47% respectively), when compared with operation on petrol (Sulaiman et al. 2013). In another study, the emissions of CO and CO_2 were found to decrease when using LPG instead of gasoline, but they also reported an increase in the NOx emissions, which they attributed to the higher flame temperature resulting in hotter engine conditions (Talib Hashem et al. 2023).

Dimethyl Ether (DME)

DME or Dimethyl ether is the simplest of the ethers, having the chemical formula C_2H_6O. DME is generally produced from syngas, gas generated from natural gas, coal, or even biomass (Tripathi and Agarwal 2023). The use of DME is beneficial in CI engines owing to its higher cetane number, indicating better knock conditions, higher latent heat leading to lower engine temperature, and higher oxygen content resulting in the formation of lesser soot, as compared to diesel (Hua 2023). However, DME is a strong solvent for elastomers and plastics used for making parts of the engine's fuel injection system. Hence, for using DME in CI engines, the susceptible components need to be produced with DME-compatible materials (Kumar and Agarwal 2021). Thomas et al. (Thomas et al. 2014) in their review of emissions from DME-fuelled CI engines have stated that the NOx and particulate matter emissions decrease with the usage of DME in CI engines as compared to diesel. However, they pointed out that the presence of ultra-fine particulate matter (UFPM) in DME emissions needs to be properly investigated as UFPM is more toxic than coarse particles.

4.6 Biofuels and Other Alternatives

Biodiesel

Biodiesel or FAME (fatty acid methyl esters) is obtained via the transesterification of triglycerides, basic fats. The sources of biodiesel may be varied, like vegetable oils (palm oil, soybean oil, rapeseed oil, etc.), algal oils, waste cooking oil, non-edible oils (jatropha, neem, etc.), and even animal fats (Kralova and Sjöblom 2010). The benefits of biodiesel include its renewable and non-toxic nature, properties nearly similar to petroleum diesel, lower pollutants in emissions (less CO, SO_2, hydrocarbons, and aromatics), and smoother engine operation (due to higher cetane rating) (Bajpai and Tyagi 2006). However, as compared to diesel, use of biodiesel has some disadvantages too, like high viscosity, lower heating value, freezing in cold conditions, lower oxidative stability, and choking of injector (Kralova and Sjöblom 2010). To put a check on the adverse effects of biodiesel, it is generally blended with petroleum diesel for commercial use. In India, the National Policy on Biofuels (NPB) adopted in 2018 set an achievable target of 5% biodiesel blending by 2030 as compared to the previous policies, which targeted 20% blending (Das 2020). In a recent review of engine performance and emission characteristics of biodiesel blends (Palani et al. 2022), the authors have stated that brake thermal efficiency decreases while the brake specific fuel consumption increases with the use of biodiesel blended fuel as compared to pure diesel. They have also noted that researches show a decrease in CO, HC, and particulate matter when diesel is replaced by biodiesel blends, alongside an increase in NOx. Biodiesel hence, is a promising biofuel for a sustainable future.

Methanol

Methanol (CH_3OH) is the simplest of the alcohol family and is an effective candidate as an alternate fuel. A review on methanol production technologies (Deka et al. 2022) has stated that among the many pathways and feedstocks to produce methanol, the highest yield is via gasification of lignocellulosic matter and catalytic reforming of the generated syngas. Methanol has some inherent issues as an engine fuel like problem in cold starting, high fuel consumption, and irregular emissions (Zhou et al. 2024b). Emission-wise, methanol-blended fuel helps reduce CO, HC, CO_2, and particulate matter in the exhaust as compared to gasoline, but NOx tends to increase (Ho et al. 2022). Another study has suggested limiting methanol blending in gasoline to up to 10% to keep the NOx and other emissions in check (Mishra et al. 2020). Methanol blending with diesel, however, can suitably reduce NOx emissions by 30% with an increase in engine performance and efficiency (Stancin et al. 2020).

Ethanol

Ethanol (C_2H_5OH) is a simple alcohol produced by fermentation of sugars obtained naturally or from sugars derived from carbohydrate or lignocellulosic biomass. The main feedstocks for ethanol production around the world are corn (60%), followed by sugarcane (25%). In 2020, the highest producers of ethanol in descending order were the United States, Brazil, Europe, China, and India (Hoang and Nghiem 2021). As per the National Biofuel Policy (NBP) of India, adopted in 2018, the target is to achieve 20% blending of ethanol in petrol by 2030. Already, the country has successfully achieved 10% blending in June 2022 (Chandel et al. 2024). Ethanol blending in gasoline implies better octane rating, better volumetric efficiency, higher output, and increase in the overall efficiency of the engine. It has been noted that there is an overall decrease in the pollutants with ethanol-blended fuels, especially HC and NOx, but the CO_2 content tends to increase in the exhaust (Chansauria and Mandloi 2018). Another review article has observed that brake power and torque increase upon each increment of ethanol percentage blending in petrol up to 20%, whereas the brake specific fuel consumption also increases with ethanol %. Brake thermal efficiency is also the maximum in the E20 blend beyond which it starts dropping (Thakur et al. 2017). As per another study (Hahn-Hägerdal et al. 2006), the ethanol produced from sugar or starch at present will not be sustainable for the future, and the extensive production of ethanol to meet the demand has to be through the conversion of lignocellulosic material. They have expressed the need of: i. Better and cheaper enzymes for enzymatic hydrolysis, ii. More active and tolerant fermenting organisms for higher productivity, and iii. Process integration for cutting down the use of energy and water, along with the reduction of generated waste.

Biogas

Biogas is generated by anaerobic digestion of wastes like cattle dung, poultry droppings, pig manure, even agricultural, food, and municipal solid wastes. The organic matter is digested by four different types of bacteria, which act in a sequence of a food chain to produce the end product, which is biogas. These bacteria, in order of their action, are: i. Hydrolytic bacteria, ii. Acidogenic bacteria, iii. Acetogenic bacteria, and iv. Methanogenic bacteria. Typical composition of biogas is mostly methane (50–75%) mixed with CO_2 (25–50%) and a small amount of H_2S (0–3%) (Arutyunov et al. 2021). On purifying biogas to attain CH_4 levels of 95–98%, which also increases the heating value from 20–30 MJ/m^3 to above 35 MJ/m^3, the biogas is now called biomethane. This biomethane has the same properties as natural gas and hence can be directly fed into the gas grid or automotive engines (Mignogna 2023). Another biogas-related technology that is currently gaining ground is compressed biogas or CBG, which has the same characteristics as CNG and can be used for cooking applications, vehicular fuel, and power generation. For producing CBG, biogas is first purified to biomethane and then compressed to around 250 bar, which increases the energy density and also aids in storage and transportation. In India, as of 2023, there are 46 CBG plants with a combined production capacity of 272 tonnes per day (Singh et al. 2023). As per a study conducted in 2021, India has the potential to generate approximately 74.8 billion cubic metres of biogas per year from its organic wastes, out of which only 3.6 billion cubic metres per year has been achieved (Singh and Kalamdhad 2023).

Producer Gas

Producer gas (PG) is generated by partial oxidation followed by reduction of biomass in specialized reactors called gasifiers. The main combustible components of PG are H_2 (12–20%), CO (17–22%), and traces of methane (2–3%) mixed with non-combustible gases like CO_2 and N_2. The quality of the gas varies depending on the feedstock used and typically lies between 5 and 5.9 MJ/Nm^3 for biomass gasification (Kalita and Baruah 2018). PG has the versatility of being used as both SI and CI engine fuel. In CI engines, PG can be utilized in dual fuel mode with diesel as pilot fuel, and up to 90% diesel replacement has been reported. On the other hand, a common SI engine can be made to run only on PG in mono fuel mode, but with minor modifications (Baruah et al. 2021). In a review of dual-fuel CI engine operation on PG (Singh et al. 2021), the authors mentioned that higher compression ratios, optimum injection timing, and optimum quantity of PG introduction into the engine greatly help improve engine performance along with emission characteristics. In another study of PG use in SI engines (Gurumurthy et al. 2020), the authors have suggested modifications in the fuel delivery system to ensure a lean air-fuel mixture, as PG has a low heating value. Furthermore, they mentioned that since the compression ratio of SI engines is limited, CI engines may be modified with an ignition system to optimally burn PG at higher pressures.

Hydrogen

Hydrogen is the lightest element and also one of the most promising alternative fuels. Hydrogen has thrice the heating value and near-zero emissions when compared to petroleum fuels (Fayaz et al. 2012). Based on parameters like source of energy, feedstock, and production process, hydrogen is assigned various "colours". For example, green hydrogen is produced by electrolysis of water using renewable energy, brown hydrogen is generated by gasification of lignite or biomass, whereas grey hydrogen is formed by steam reforming of natural gas using non-renewable energy (Incer-Valverde et al. 2023). Hydrogen engines are 25% more efficient than their gasoline counterpart, besides having lower emissions, but there is a problem of storing hydrogen in vehicle tanks to ensure sufficient driving range (Vinoth Kanna and Paturu 2020). Hydrogen is highly suitable for engines due to high flame speed, low ignition energy, and high adiabatic temperature, but the only drawback is increased NOx due to hotter engine conditions (Falfari et al. 2023). Researches have indicated an increase of hydrogen requirement by eight times of current demand. They have suggested the utilization of solar PV, hydropower, and offshore wind turbines as energy options for the generation of green hydrogen to meet the requirements of the future (Saha et al. 2024).

4.7 Advantages of Alternative Fuels

The above discussion enumerating the different types of alternative fuels that are currently in use or under research already highlights the key advantages over conventional fossil fuels. However, to present a comprehensive description of performance or emissions, Table 4.5 enlists the popularly used alternative fuels again with their respective features, which make them more suitable than conventional fuels.

4.8 Conclusions

This chapter provides an overview of liquid and gaseous fuels, alternative fossil fuels, and biofuels, emphasizing their properties, applications, and future potential for internal combustion (IC) engines. The chapter highlights the critical properties of fuels, such as gross calorific value, octane/cetane numbers, density, and combustion characteristics, which determine their suitability for spark-ignition (SI) and compression-ignition (CI) engines. While SI engines require high-octane fuels for smooth combustion and resistance to knocking, CI engines demand high-cetane fuels for efficient ignition under high-pressure conditions.

Conventional liquid fuels like gasoline, diesel, and kerosene continue to dominate due to their energy density and infrastructure, while gaseous fuels such as natural gas, propane, and hydrogen are gaining prominence for their cleaner combustion and lower emissions. Hydrogen, with its zero-emission profile and high energy content, is particularly promising for sustainable applications.

Alternative fossil fuels, including compressed natural gas (CNG), liquefied natural gas (LNG), liquefied petroleum gas (LPG), and dimethyl ether (DME), offer reduced emissions and better engine performance but require advancements in infrastructure and handling systems.

Biofuels, such as biodiesel, ethanol, and biogas, stand out for their renewability, reduced greenhouse gas emissions, and compatibility with existing IC engine technologies. Hydrogen-based fuels, especially green hydrogen, represent a transformative solution but face challenges in production and storage.

In conclusion, transitioning from conventional fuels to alternative and biofuels is essential for improving IC engine efficiency, reducing environmental impact, and ensuring energy security. Continued innovation, coupled with supportive policies and infrastructure, will be vital to unlocking the full potential of these advanced fuel technologies, paving the way for sustainable IC engine solutions.

Table 4.5 Performance and emission characteristics of alternative fuels

Alternative fuel	Engine performance	Emissions
Compressed natural gas (CNG)	Compared to gasoline (Jahirul et al. 2010): • Lower BP (brake power) and lower BSFC (brake specific fuel consumption) • Slightly higher BTE (brake thermal efficiency) When the amount of CNG is increased in diesel-run CI engine (Jamrozik et al. 2019): • Ignition delay increases • Maximum pressure increases • Most stable at a 30–45% share of CNG	Compared to gasoline (Jahirul et al. 2010): • Higher EGT (Exhaust Gas temperature) • Higher NO_x • Other emissions like HC, CO, and CO_2 lower When the amount of CNG is increased in diesel-run CI engine (Jamrozik et al. 2019): • NO_x increases • HC reduces • CO_2 reduces • CO almost nullified
Liquified petroleum gas (LPG) Simsek and Uslu (2020)	When compared to gasoline in an SI engine: • BSFC increases • BTE decreases • Cylinder pressure decreases	When compared to gasoline in an SI engine: • HC decreases • CO_2 decreases • CO reduced
Biodiesel Altarazi et al. (2022)	When biodiesel is blended with diesel for CI engine: • BTE decreases • BSFC increases	When biodiesel is blended with diesel for CI engine: • HC decreases • NOx increases
Ethanol Mohammed et al. (2021)	Increase of ethanol % in blend with gasoline: • Increase in knock resistance • Fuel efficiency can be improved by downsizing, downspeeding, increasing compression ratio (CR), etc	Increase of ethanol % in blend with gasoline: • Decrease emissions of 1,3-butadiene and benzene • lower PM (particulate matter) emissions • CO and NOx emissions are not affected significantly
Biogas Vasan et al. (2024)	Biogas usage in IC engines results in: • Increased energy conversion efficiency • Higher fuel efficiency	Biogas usage in IC engines helps in: • Lowering NOx emissions • Reduced CO and HC emissions

(continued)

Table 4.5 (continued)

Alternative fuel	Engine performance	Emissions
Producer gas Raj et al. (2024)	• Lower BP compared to diesel • Lower thermal efficiency • Better suitability in dual-fuel mode with diesel for improved performance	• Higher CO_2 emissions compared to CNG but lower than diesel • NOx emissions are slightly higher compared to diesel • HC and CO emissions reduced
Hydrogen Algayyim et al. (2024)	• Higher thermal efficiency • Higher flame speed leading to increased power output • Can operate with lean mixtures, improving fuel economy and efficiency	• Zero CO and HC emissions due to the absence of carbon • NOx emissions increase due to higher combustion temperatures • No particulate matter (PM) emissions

References

Abubakar S, Muhamad Said MF, Abas MA, Ismail NA, Khalid AH, Roslan MF et al (2024) Hydrogen-fuelled internal combustion engines—bibliometric analysis on research trends, hotspots, and challenges. Int J Hydrogen Energy 61:623–638. https://doi.org/10.1016/j.ijhydene.2024.02.280

Algayyim SJM, Saleh K, Wandel AP, Fattah IMR, Yusaf T, Alrazen HA (2024) Influence of natural gas and hydrogen properties on internal combustion engine performance, combustion, and emissions: a review. Fuel 362:130844. https://doi.org/10.1016/j.fuel.2023.130844

Ali Abuelnour M, Farid Muhamad Said M, Abuelnuor AAA (2024) Numerical investigation of hydrogen-diesel dual fuel engine: combustion and emissions characteristics under varying hydrogen concentration and exhaust gas recirculation. Fuel 377:132794. https://doi.org/10.1016/J.FUEL.2024.132794

Altarazi YSM, Abu Talib AR, Yu J, Gires E, Abdul Ghafir MF, Lucas J et al (2022) Effects of biofuel on engines performance and emission characteristics: a review. Energy 238:121910. https://doi.org/10.1016/J.ENERGY.2021.121910

Arefin MA, Nabi MN, Akram MW, Islam MT, Chowdhury MW (2020) A review on liquefied natural gas as fuels for dual fuel engines: opportunities, challenges and responses. Energies 13. https://doi.org/10.3390/en13226127

Arutyunov V, Nikitin A, Strekova L, Savchenko V, Sedov I (2021) Utilization of renewable sources of biogas for small-scale production of liquid fuels. Catal Today 379:23–27. https://doi.org/10.1016/j.cattod.2020.06.057

Askari O, Elia M, Ferrari M, Metghalchi H (2017) Auto-ignition characteristics study of gas-to-liquid fuel at high pressures and low temperatures. J Energy Resour Technol Trans ASME 139:1–6. https://doi.org/10.1115/1.4033983

Astbury GR (2008) A review of the properties and hazards of some alternative fuels. Process Saf Environ Prot 86:397–414. https://doi.org/10.1016/j.psep.2008.05.001

Awad OI, Ali OM, Mamat R, Abdullah AA, Najafi G, Kamarulzaman MK et al (2017) Using fusel oil as a blend in gasoline to improve SI engine efficiencies: a comprehensive review. Renew Sustain Energy Rev 69:1232–1242. https://doi.org/10.1016/j.rser.2016.11.244

Awad OI, Mamat R, Ali OM, Sidik NAC, Yusaf T, Kadirgama K et al (2018) Alcohol and ether as alternative fuels in spark ignition engine: a review. Renew Sustain Energy Rev 82:2586–2605. https://doi.org/10.1016/j.rser.2017.09.074

Bae C, Kim J (2017) Alternative fuels for internal combustion engines. Proc Combust Inst 36:3389–3413. https://doi.org/10.1016/j.proci.2016.09.009

Bajpai D, Tyagi VK (2006) Biodiesel: source, production, composition, properties and its benefits. J Oleo Sci 55:487–502. https://doi.org/10.5650/jos.55.487

Baruah D, Kalita P, Moholkar VS (2021) A Comprehensive study on utilization of producer gas as IC engine fuel. In: Singh AP, Kumar D, Agarwal AK, (eds) Alternative fuels and advanced combustion techniques as sustain solutions internal combustion engines. Springer, pp 117–47. https://doi.org/10.1007/978-981-16-1513-9_6

Bhowmik S, Panua R, Ghosh SK (2024) Investigation of performance, combustion and exhaust emission characteristics of a compression ignition engine fuelled with diesel-kerosene-ethanol-hydrogen strategies. Int J Hydrogen Energy 49:697–712. https://doi.org/10.1016/J.IJHYDENE.2023.09.035

Caton JA (2010) Implications of fuel selection for an SI engine: results from the first and second laws of thermodynamics. Fuel 89:3157–3166. https://doi.org/10.1016/j.fuel.2010.05.005

Chandel AK, Arora R, Gupta R, Narisetty V, Kumar V (2024) Global biofuel alliance and roadmap for ethanol blending program in India: analysis and perspective. Sugar Tech 26:1005–1015. https://doi.org/10.1007/s12355-024-01470-5

Chansauria P, Mandloi RK (2018) Effects of ethanol blends on performance of spark ignition engine-a review. Mater Today Proc 5:4066–4077. https://doi.org/10.1016/j.matpr.2017.11.668

Choi MS (2011) LNG for petroleum engineers. SPE Proj Facil Constr 6:255–263. https://doi.org/10.2118/133722-PA

Civan F (1989) Review of methods for measurement of natural gas specific gravity. Soc Pet Eng 173–186. https://doi.org/10.2523/19073-ms

Correa Pabón RE, Souza Filho CR, de. (2019) Crude oil spectral signatures and empirical models to derive API gravity. Fuel 237:1119–1131. https://doi.org/10.1016/j.fuel.2018.09.098

Coutinho JAP, Mirante F, Ribeiro JC, Sansot JM, Daridon JL (2002) Cloud and pour points in fuel blends. Fuel 81:963–967. https://doi.org/10.1016/S0016-2361(01)00213-7

Dagaut P, Cathonnet M (2006) The ignition, oxidation, and combustion of kerosene: a review of experimental and kinetic modeling. Prog Energy Combust Sci 32:48–92. https://doi.org/10.1016/J.PECS.2005.10.003

Das S (2020) The national policy of biofuels of India—a perspective. Energy Policy 143:111595. https://doi.org/10.1016/j.enpol.2020.111595

Das AK, Sahu SK, Panda AK (2022) Current status and prospects of alternate liquid transportation fuels in compression ignition engines: a critical review. Renew Sustain Energy Rev 161:112358. https://doi.org/10.1016/j.rser.2022.112358

Deka TJ, Osman AI, Baruah DC, Rooney DW (2022) Methanol fuel production, utilization, and techno-economy: a review. Environ Chem Lett 20:3525–3554. https://doi.org/10.1007/s10311-022-01485-y

Dembicki Jr H (2017) Interpreting crude oil and natural gas data. Practical petroleum geochemistry for exploration and production. Elsevier, pp 135–88. https://doi.org/10.1016/b978-0-12-803350-0.00004-0

Demirbas A, Balubaid MA, Basahel AM, Ahmad W, Sheikh MH (2015) Octane rating of gasoline and octane booster additives. Pet Sci Technol 33:1190–1197. https://doi.org/10.1080/10916466.2015.1050506

EIA. International Energy Outlook (2016)

Eyankware O, Ulakpa C, Eyankware MO, Author C (2016) Determination of cloud and pour point of crude oil with reference to crude transportation. Int J Sci Healthc Res 1:20–28

Falfari S, Cazzoli G, Mariani V, Bianchi GM (2023) Hydrogen application as a fuel in internal combustion engines. Energies 16. https://doi.org/10.3390/en16062545

Fayaz H, Saidur R, Razali N, Anuar FS, Saleman AR, Islam MR (2012) An overview of hydrogen as a vehicle fuel. Renew Sustain Energy Rev 16:5511–5528. https://doi.org/10.1016/j.rser.2012.06.012

Gad MS, Ismail MA (2021) Effect of waste cooking oil biodiesel blending with gasoline and kerosene on diesel engine performance, emissions and combustion characteristics. Process Saf Environ Prot 149:1–10. https://doi.org/10.1016/J.PSEP.2020.10.040

Ge JC, Luo H, Choi NJ (2024) Investigation of bioethanol low-carbon fuel for diesel engines under idling conditions: combustion, engine performance and emissions. Green Energy Resour 100100. https://doi.org/10.1016/J.GERR.2024.100100

Ghetti P (1994) A rapid heating TGA method for evaluating the carbon residue of fuel oil. Fuel 73:1918–1921. https://doi.org/10.1016/0016-2361(94)90222-4

Ghosh S, Majumder R, Chatterjee B (2024) Natural gas matters: LNG and India's quest for clean energy. Gases 4:1–17. https://doi.org/10.3390/gases4010001

Gültekin N (2024) Experimental study of the effects of diesel, bioethanol, and hydrogen on combustion, emissions, mechanical vibration, and noise in a CI engine with different valve lift. Int J Hydrogen Energy 93:1011–1021. https://doi.org/10.1016/J.IJHYDENE.2024.11.012

Gurumurthy V, Babu U, Kumararaja L (2020) Performance of SI engine fuelled with producer gas at various spark timings and compression ratios-A review. AIP Conf Proc 2225. https://doi.org/10.1063/5.0005828

Hahn-Hägerdal B, Galbe M, Gorwa-Grauslund MF, Lidén G, Zacchi G (2006) Bio-ethanol—the fuel of tomorrow from the residues of today. Trends Biotechnol 24:549–556. https://doi.org/10.1016/j.tibtech.2006.10.004

Han WQ, De YC (2015) Research on high cetane and high octane number fuels and the mechanism for their common oxidation and auto-ignition. Fuel 150:29–40. https://doi.org/10.1016/j.fuel.2015.01.090

Ho CS, Peng J, Yun UH, Zhang Q, Mao H (2022) Impacts of methanol fuel on vehicular emissions: a review. Front Environ Sci Eng 16:1–20. https://doi.org/10.1007/s11783-022-1553-4

Hoang TD, Nghiem N (2021) Recent developments and current status of commercial production of fuel ethanol. Fermentation 7. https://doi.org/10.3390/fermentation7040314

Hua Y (2023) Ethers and esters as alternative fuels for internal combustion engine: a review. Int J Engine Res 24:178–216. https://doi.org/10.1177/14680874211046480

Hua Y (2024) Research progress of higher alcohols as alternative fuels for compression ignition engines. Fuel 357:129749. https://doi.org/10.1016/j.fuel.2023.129749

Ilhak MI, Tangoz S, Akansu SO, Kahraman N (2019) Alternative fuels for internal combustion engines. The future internal combustion engines. Intech Open. https://doi.org/10.5772/intechopen.76548

Incer-Valverde J, Korayem A, Tsatsaronis G, Morosuk T (2023) "Colors" of hydrogen: definitions and carbon intensity. Energy Convers Manag 291:117294. https://doi.org/10.1016/j.enconman.2023.117294

Inspire SGS. Methanol : Properties (2020)

Ito E, van Veen JAR (2006) On novel processes for removing sulphur from refinery streams. Catal Today 116:446–460. https://doi.org/10.1016/j.cattod.2006.06.040

Jahirul MI, Masjuki HH, Saidur R, Kalam MA, Jayed MH, Wazed MA (2010) Comparative engine performance and emission analysis of CNG and gasoline in a retrofitted car engine. Appl Therm Eng 30:2219–2226. https://doi.org/10.1016/J.APPLTHERMALENG.2010.05.037

Jain S, Kumar S (2024) A comprehensive review of bioethanol production from diverse feedstocks: current advancements and economic perspectives. Energy 296:131130. https://doi.org/10.1016/j.energy.2024.131130

Jamrozik A, Tutak W, Grab-Rogaliński K (2019) An experimental study on the performance and emission of the diesel/CNG dual-fuel combustion mode in a stationary CI engine. Energies 12. https://doi.org/10.3390/en12203857.

Johnson C, Mashuga CV (2023) Reconsidering autoignition in the context of modern process safety: literature review and experimental analysis. J Loss Prev Process Ind 81:104963. https://doi.org/10.1016/j.jlp.2022.104963

Kalita P, Baruah D (2018) Investigation of biomass gasifier product gas composition and its characterization. In: De S, Agarwal AK, Moholkar VS, Bhaskar T (eds) Coal biomass gasification:

recent advances and future challenges. Springer Singapore, pp 115–49. https://doi.org/10.1007/978-981-10-7335-9_5
Kar SK, Sinha ASK, Harichandan S, Bansal R, Balathanigaimani MS (2023) Hydrogen economy in India: a status review. Wiley Interdiscip Rev Energy Environ 12:e459. https://doi.org/10.1002/wene.459
Gad SC (2014) Kerosene. Encyclopedia of toxicology, Fourth Ed. Academic Press, vol 1–9, vol 5, pp V5-747-V5-750. https://doi.org/10.1016/B978-0-12-824315-2.00809-5
Kiang Y-H (2018) Fuel property estimation and combustion process characterization. Academic Press
Kontses A, Triantafyllopoulos G, Ntziachristos L, Samaras Z (2020) Particle number (PN) emissions from gasoline, diesel, LPG, CNG and hybrid-electric light-duty vehicles under real-world driving conditions. Atmos Environ 222:117126. https://doi.org/10.1016/j.atmosenv.2019.117126
Kralova I, Sjöblom J (2010) Biofuels-renewable energy sources: a review. J Dispers Sci Technol 31:409–425. https://doi.org/10.1080/01932690903119674
Kumar V, Agarwal AK (2021) Material compatibility, technical challenges and modifications required for DME adaptation in compression ignition engines. In: Singh AS, Kumar D, Agarwal AK (eds) Alternative fuels advanced combustion techniques as sustainable solutions for internal combustion engines, pp 37–57. https://doi.org/10.1007/978-981-16-1513-9_3
Lee CC, Tran MV, Tan BT, Scribano G, Chong CT (2021) A comprehensive review on the effects of additives on fundamental combustion characteristics and pollutant formation of biodiesel and ethanol. Fuel 288:119749. https://doi.org/10.1016/j.fuel.2020.119749
Leonardo A, Semin (2022) Effect of CNG engine conversion on performance characteristic: a review. IOP Conf Ser Earth Environ Sci 972. https://doi.org/10.1088/1755-1315/972/1/012028
Li Y, Huang Y, Chen H, Wei F, Zhang Z, Zhou M (2024a) Combustion and emission of diesel/PODE/gasoline blended fuel in a diesel engine that meet the China VI emission standards. Energy 301:131473. https://doi.org/10.1016/J.ENERGY.2024.131473
Li G, Wang J, Li K, Yang K, Han W, Li S (2024b) Effects of isopropanol-butanol-ethanol on the performance, combustion and emission characteristics of a diesel-methanol dual-fuel engine. Renew Energy 237:121790. https://doi.org/10.1016/J.RENENE.2024.121790
Livaniou S, Papadopoulos GA (2022) Liquefied natural gas (LNG) as a transitional choice replacing marine conventional fuels (heavy fuel oil/marine diesel oil), towards the era of decarbonisation. Sustainability 14. https://doi.org/10.3390/su142416364
Ljubas D, Krpan H, Matanoviae I (2010) Influence of engine oils dilution by fuels on their viscosity, flash point and fire point. Nafta 61:73–79
Ma Q, Wang J, Jin S, Shi M (2024) The impact of ignition and activation energy distribution on the combustion and emission characteristics of diesel-ammonia-natural gas engines. J Energy Inst 101903. https://doi.org/10.1016/j.joei.2024.101903
Mignogna D, Ceci P, Cafaro C, Corazzi G, Avino P (2023) Production of biogas and biomethane as renewable energy sources: a review. Appl Sci 13. https://doi.org/10.3390/app131810219
Mishra PC, Gupta A, Kumar A, Bose A (2020) Methanol and petrol blended alternate fuel for future sustainable engine: a performance and emission analysis. Meas J Int Meas Confed 155:107519. https://doi.org/10.1016/j.measurement.2020.107519
Mohammed MK, Balla HH, Al-Dulaimi ZMH, Kareem ZS, Al-Zuhairy MS (2021) Effect of ethanol-gasoline blends on SI engine performance and emissions. Case Stud Therm Eng 25:100891. https://doi.org/10.1016/J.CSITE.2021.100891
Nyongesa AJ, Choi JH, Lee JW, Lee J, Kim S, Yeo S, et al (2024) Effects of natural gas admission location and timing on performance and emissions characteristics of LPDF two-stroke engine at low load. Case Stud Therm Eng 56:104241. https://doi.org/10.1016/J.CSITE.2024.104241.
Oasmaa A, Leppämäki E, Koponen P, Levander J, Tapola E (1997) Physical characterisation of biomass-based pyrolysis liquids application of standard fuel oil analyses. VTT Publ. https://doi.org/10.1016/s0140-6701(98)97220-4

Oktem A, Oktem AG (2024) Bioethanol obtained from soluble sugars in sweet sorghum, influencing properties and their interrelationships. Ind Crops Prod 221:119351. https://doi.org/10.1016/j.indcrop.2024.119351

Palani Y, Devarajan C, Manickam D, Thanikodi S (2022) Performance and emission characteristics of biodiesel-blend in diesel engine: a review. Environ Eng Res 27:0–1. https://doi.org/10.4491/eer.2020.338

Patil V, Singh P, Kumar S (2024) Reliability of renewable oxygenated fuels in engines: a comprehensive review of long-term testing. Fuel 367:131451. https://doi.org/10.1016/j.fuel.2024.131451

de Paulo EH, dos Santos FD, Folli GS, Santos LP, Nascimento MHC, Moro MK et al (2022) Determination of gross calorific value in crude oil by variable selection methods applied to 13C NMR spectroscopy. Fuel 311:0–3. https://doi.org/10.1016/j.fuel.2021.122527

Paykani A, Chehrmonavari H, Tsolakis A, Alger T, Northrop WF, Reitz RD (2022) Synthesis gas as a fuel for internal combustion engines in transportation. Prog Energy Combust Sci 90:100995. https://doi.org/10.1016/j.pecs.2022.100995

Pham Q, Park S, Agarwal AK, Park S (2022) Review of dual-fuel combustion in the compression-ignition engine: spray, combustion, and emission. Energy 250:123778. https://doi.org/10.1016/j.energy.2022.123778

Pourkhesalian AM, Shamekhi AH, Salimi F (2010) Alternative fuel and gasoline in an SI engine: a comparative study of performance and emissions characteristics. Fuel 89:1056–1063. https://doi.org/10.1016/j.fuel.2009.11.025

Raj R, Tirkey JV, Jena P, Prajapati LK (2024) Comparative analysis of Gasifier-CI engine performance and emissions characteristics using diesel with producer gas derived from coal– briquette-coconut shell-mahua feedstock and its blends. Energy 293:130708. https://doi.org/10.1016/J.ENERGY.2024.130708

Rathore NS, Panwar NL, Chiplunkar YV (2009) Design and techno economic evaluation of biomass gasifier for industrial thermal applications. African J Environ Sci Technol 3:006–012. https://doi.org/10.5897/AJEST08.102

Reddy P, Letcher TM (2007) Phase equilibrium studies on ionic liquid systems for industrial separation processes of complex organic mixtures. In: Thermodynamics, Solubility and Environmental Issues. Elsevier, pp 85–111. https://doi.org/10.1016/B978-044452707-3/50007-0

Roman FF, Díaz De Tuesta JL, Silva AMT, Faria JL, Gomes HT (2021) Carbon-based materials for oxidative desulfurization and denitrogenation of fuels: a review. Catalysts 11:1–41. https://doi.org/10.3390/catal11101239

Saha P, Akash FA, Shovon SM, Monir MU, Ahmed MT, Khan MFH et al (2024) Grey, blue, and green hydrogen: a comprehensive review of production methods and prospects for zero-emission energy. Int J Green Energy 21:1383–1397. https://doi.org/10.1080/15435075.2023.2244583

Sahu TK, Shukla PC (2023) Combustion and emission characteristics of oxygenated alternative fuels in compression ignition engines. In: Shukla PC, Belgiorno G, Blasio GD, Agarwal AK (eds) Renewable fuels for sustainable mobility. Energy, Environment Sustainability. Springer, Singapore, pp 79–95. https://doi.org/10.1007/978-981-99-1392-3_4

Search C (2018) Combustion of fuels—carbon dioxide emission environmental emission of carbon dioxide CO_2 when combustion fuels like coal, oil, natural gas, LPG and bio energy sorry to see that you are blocking ads on The Engineering ToolBox ! Disabling your adblo, pp 1–7

Sedghi R, Shahbeik H, Rastegari H, Rafiee S, Peng W, Nizami AS et al (2022) Turning biodiesel glycerol into oxygenated fuel additives and their effects on the behavior of internal combustion engines: a comprehensive systematic review. Renew Sustain Energy Rev 167:112805. https://doi.org/10.1016/j.rser.2022.112805

Semin RAB (2008) A technical review of compressed natural gas as an alternative fuel for internal combustion engines semin. Rosli Abu Bakar Automotive Excellent Center , Faculty of Mechanical Engineering. Am J Eng Appl Sci 1:302–11

Shah R, Schevon A (2022) Recent advances in the realm of cloud point and pour point testing specifically as it relates to biofuels. Fuel

Shirneshan A, Kanberoglu B, Gonca G (2025) Experimental investigation and parametric modeling of the effect of alcohol addition on the performance and emissions characteristics of a diesel engine fueled with biodiesel-diesel-hydrogen fuel mixtures. Fuel 381:133489. https://doi.org/10.1016/J.FUEL.2024.133489
Simsek S, Uslu S (2020) Investigation of the impacts of gasoline, biogas and LPG fuels on engine performance and exhaust emissions in different throttle positions on SI engine. Fuel 279:118528. https://doi.org/10.1016/J.FUEL.2020.118528
Singh P, Kalamdhad AS (2023) A comprehensive assessment of state-wise biogas potential and its utilization in India. Biomass Convers Biorefinery 13:12557–12579. https://doi.org/10.1007/s13399-021-02001-y
Singh P, Kumar R, Sharma S, Kumar S (2021) Effect of engine parameters on the performance of dual-fuel CI engines with producer gas—a review. Energy Fuels 35:16377–16402. https://doi.org/10.1021/acs.energyfuels.1c02279
Singh D, Tembhare M, Kumar S (2023) Compressed biogas plants in India: existing status, technological advances and challenges. Biomass Bioenerg 178:106964. https://doi.org/10.1016/j.biombioe.2023.106964
Singh MK, Chaudhary VP, Lata DB (2024) Experimental investigations on the performance and emission characteristics of hydrogen enriched Bio-CNG in a common rail direct injection dual fuel diesel engine. Int J Hydrogen Energy 88:579–593. https://doi.org/10.1016/J.IJHYDENE.2024.09.236
Sivaramakrishnan K, Ravikumar P (2012) Determination of cetane number of biodiesel and it's influence on physical properties. ARPN J Eng Appl Sci 7:205–211
Stančin H, Mikulčić H, Wang X, Duić N (2020) A review on alternative fuels in future energy system. Renew Sustain Energy Rev 128. https://doi.org/10.1016/j.rser.2020.109927
Sulaiman MY, Ayob MR, Meran I (2013) Performance of single cylinder spark ignition engine fueled by LPG. Procedia Eng 53:579–585. https://doi.org/10.1016/j.proeng.2013.02.074
Tahir MM, Ali MS, Salim MA, Bakar RA, Fudhail AM, Hassan MZ et al (2015) Performance analysis of a spark ignition engine using compressed natural gas (CNG) as fuel. Energy Procedia 68:355–362. https://doi.org/10.1016/j.egypro.2015.03.266
Talib Hashem G, Al-Dawody MF, Sarris IE (2023) The characteristics of gasoline engines with the use of LPG: an experimental and numerical study. Int J Thermofluids 18:100316. https://doi.org/10.1016/j.ijft.2023.100316
Thakur AK, Kaviti AK, Mehra R, Mer KKS (2017) Performance analysis of ethanol–gasoline blends on a spark ignition engine: a review. Biofuels 8:91–112. https://doi.org/10.1080/17597269.2016.1204586
Thomas G, Feng B, Veeraragavan A, Cleary MJ, Drinnan N (2014) Emissions from DME combustion in diesel engines and their implications on meeting future emission norms: a review. Fuel Process Technol 119:286–304. https://doi.org/10.1016/j.fuproc.2013.10.018
Thomson SM, Robertson G, Myers R, Schütze A (2003) Liquefied petroleum gas. In: Elvers B, Schütze A (eds) Handbook of fuels energy sources for transportation 2nd ed, Wiley-VCH, pp 101–117. https://doi.org/10.1016/B978-075064452-5/50069-9
Tripathi A, Agarwal AK (2023) DME as a green fuel for transport sector, pp 35–77. https://doi.org/10.1007/978-981-99-1392-3_3
Ugurlu A, Oztuna S (2015) A comparative analysis study of alternative energy sources for automobiles. Int J Hydrogen Energy 40:11178–11188. https://doi.org/10.1016/j.ijhydene.2015.02.115
Vasan V, Sridharan NV, Feroskhan M, Vaithiyanathan S, Subramanian B, Tsai PC et al (2024) Biogas production and its utilization in internal combustion engines—a review. Process Saf Environ Prot 186:518–539. https://doi.org/10.1016/J.PSEP.2024.04.014
Verma TN, Shrivastava P, Rajak U, Dwivedi G, Jain S, Zare A et al (2021) A comprehensive review of the influence of physicochemical properties of biodiesel on combustion characteristics, engine performance and emissions. J Traffic Transp Eng (English Ed) 8:510–533. https://doi.org/10.1016/j.jtte.2021.04.006

Vinoth Kanna I, Paturu P (2020) A study of hydrogen as an alternative fuel. Int J Ambient Energy 41:1433–1436. https://doi.org/10.1080/01430750.2018.1484803

Wan Ghazali WNM, Mamat R, Masjuki HH, Najafi G (2015) Effects of biodiesel from different feedstocks on engine performance and emissions: a review. Renew Sustain Energy Rev 51:585–602. https://doi.org/10.1016/j.rser.2015.06.031

Wang T, Zhou Y, Luo Z, Wen H, Zhao J, Su B et al (2020) Flammability limit behavior of methane with the addition of gaseous fuel at various relative humidities. Process Saf Environ Prot 140:178–189. https://doi.org/10.1016/j.psep.2020.05.005

Wu Y, Hu J, Lin Y, Chen P, Chen G, Wang Z (2024) Experimental study of ammonia energy ratio on combustion and emissions from ammonia-gasoline dual-fuel engine at various load conditions. J Energy Inst 117:101868. https://doi.org/10.1016/J.JOEI.2024.101868

Yakin A, Gülcan M, Çelebi S, Demir U, Yilmaz E (2024) Investigation of the performance and emission effects of ammonia-borane as a B-N-based amine-borane adduct in gasoline engines. Int J Hydrogen Energy 87:1129–1139. https://doi.org/10.1016/J.IJHYDENE.2024.09.117

Yasar A, Haider R, Tabinda AB, Kausar F, Khan M (2013) Short communication a comparison of engine emissions from heavy, medium, and light vehicles for CNG, diesel, and gasoline fuels. Polish J Environ Stud 22:1277–1281

Zareei J, widatalla STH, Guerrero JWG, Shavkatov N, Al-amir QR (2024) Dataset of numerical study of the impact of two-step injection (direct and port) of hydrogen-enriched and natural gas mixtures on engine efficiency and exhaust emissions. Data Br 57:110894. https://doi.org/10.1016/J.DIB.2024.110894

Zhou W, Wang H, Gao J, Zhou S (2024a) Effects of using argon, hydrogen and syngas on performance and emissions of a heavy-duty natural gas/diesel RCCI engine at low load. Int J Hydrogen Energy 77:958–974. https://doi.org/10.1016/J.IJHYDENE.2024.06.255

Zhou F, Yu J, Wu C, Fu J, Liu J, Duan X (2024) Science of the total environment the application prospect and challenge of the alternative methanol fuel in the internal combustion engine. Sci Total Environ 913:169708. https://doi.org/10.1016/j.scitotenv.2023.169708

Part II
Fundamentals of Ammonia Combustion

Chapter 5
Insights into Fundamentals of Ammonia Combustion

Avinash Kumar Agarwal and Dhruv Raj Karana

Abstract The transportation sector is heavily dependent on internal combustion engines and will continue to remain so. However, the fuels would undergo transition. Depletion of petroleum reserves and the paradigm shift towards decarbonization has turned ammonia into a lucrative fuel for internal combustion engines. Carbon free nature, high energy density, low cost, presence of existing supply infrastructure, etc. are highlighting features that make ammonia a superior fuel. However, like any other fuel, it too has certain disadvantages that need to be addressed. High nitrogen oxide emissions and unburnt ammonia are the two broad challenges for ammonia as a fuel which can be overcome by regulating the combustion process. Thus, there is need to understand the fundamentals of ammonia combustion in detail and fine-tune the combustion process for widespread use. This chapter would help in understanding details of the ammonia combustion/oxidation process that would in turn help in achieving greenhouse gas emission targets in the near future.

Keywords Combustion · Ammonia · Chemical kinetics · Reaction mechanisms

5.1 Introduction

Combustion has been the primary mode of energy generation for the modern society. The combustion process has an inherent character of producing some undesirable products like CO_2 which contribute greatly to the greenhouse effect on earth. These greenhouse gases are largely responsible for environmental degradation causing climate change. Mitigation of such harmful substances has led to global awareness about emissions from the combustion process. Stricter emission norms with traditional carbon fuels have led to the search of alternative fuels devoid of carbon. Ammonia is a one such fuel which possesses advantages like high energy density, high hydrogen content, presence of existing supply infrastructure, easy storage and

A. K. Agarwal (✉) · D. R. Karana
Engine Research Laboratory, Department of Mechanical Engineering, IIT Kanpur, Kanpur, India
e-mail: akag@iitk.ac.in

© The Author(s), under exclusive license to Springer Nature Singapore Pte Ltd. 2026
H. Valera and A. K. Agarwal (eds.), *Ammonia and Low Carbon Combustion Engine Technologies for De-Fossilizing Transport Sector*, Energy, Environment, and Sustainability, https://doi.org/10.1007/978-981-96-6628-7_5

transportation, etc. Currently ammonia is produced from petroleum derivatives due to economic feasibility but, it can be readily produced from renewable means making it a sustainable future fuel.

Burning of ammonia as a fuel dates to 1943, (Koch 1945) when it was used as a fuel for buses during the second world war due to disruption in the supply of petroleum. However direct burning of ammonia is a challenge due to its high ignition energy (8 mJ at equivalence ratio of 0.9) (Verkamp et al. 1967), slower flame speed and excessive NOx emissions. But the positive traits of ammonia have led to resurgence of interest amongst the industrial and academic community. Use of ammonia for maritime applications is maturing (UNCTAD 2019) and the shipping industry has already started to see it as a better alternative. The shipping sector is seriously considering ammonia to cut its carbon emissions to meet the future emission norms. Currently, the ship industry is largely dependent on diesel fuel. The soot-NOx dilemma is the major barrier in the diesel engines (Agarwal et al. 2011). Researchers (Agarwal et al. 2017; Prasad and Agarwal 2021) have already being trying to shift to advanced combustion technologies which emit very little emissions. Sustainable fuels like alcohols and biodiesels (Agarwal 2007; Agarwal et al. 2022) have also been proposed for long but complete elimination of CO_2 seems largely impossible.

Table 5.1 provides the combustion data of the commonly used fuels in comparison to ammonia. It is evident from Table 5.1 that ammonia is not the readily ignitable fuel when compared to other fuels. But its carbon free nature, easy availability makes economic sense to make it the fuel for future. As a result, alternate methods for cleaner combustion of ammonia are being searched. This has led to the importance of understanding the combustion chemistry. With the detail in understanding of the reaction kinetics of ammonia, efficient combustion system can be designed. This chapter would provide a pathway to get a brief understanding of reaction fundamentals of ammonia combustion.

Table 5.1 Combustion-related properties of ammonia and other common fuels (Kobayashi et al. 2019a)

Property	H_2 content (wt%)	Auto ignition temperature (K)	Calorific value (MJ/kg)	Laminar burning velocity (m/sec)	Stoichiometric equivalence ratio (–)
NH_3	17.7	924	18.6	0.07	6
H_2	100	844	120	2.91	34
Diesel	14	503	45.4	1.1	14.4
Gasoline	15.8	573	45	0.45	14.7
Natural gas	< 20	723	38.1	1.2	9.5

5.2 Fundamentals of Ammonia Combustion

5.2.1 Ammonia Premixed Flames

The orange flame shown in Fig. 5.1 is a NH_3-air laminar premixed flame. The flame appears orange because of combination of NH_2 α band and water vapor spectrum close to 543–665 nm. (Hayakawa et al. 2015a; Pearse and Gaydon 1976). The flame is described by certain basic parameters. One such parameters that defines flame is the laminar burning velocity. It describes how a flame propagates towards the unburnt fuel–air mixture and consumes the entire charge at constant temperature and pressure conditions. It is generally denoted by the letter S and a subscript, where the subscript denotes the conditions of measurement. Mathematically, it is calculated by using the correlation $S_u = S_{u0}\left(\frac{T_u}{T_0}\right)^{\alpha}\left(\frac{p_u}{p_0}\right)^{\beta}$ given by Metghalchi and Kech (1982). Here the subscript 'u' stands for unburnt and '0' stands for room temperature and pressure conditions. The α and β are constants. It is dependent on the unburnt mixture temperature and pressure.

The laminar flame velocity is dependent on equivalence ratio, ambient pressure and ambient temperature. Experimental studies are done to characterise the NH_3-air flames for different equivalence ratios. The effect of varying equivalence ratio on the unstretched laminar burning velocity at room temperature and atmospheric pressure is studied by many. The highest laminar burning velocity is observed at $\varphi = 1.1$, which is a marginally rich mixture. The maximum velocity observed is 7 cm/sec, which is significantly less than that of other fuel–air mixtures like methane-air, whose maximum velocity is around 35 cm/sec (Law 2006). The practical significance of laminar burning velocity is that it indicates how fast a fuel will burn when subjected to right temperature and pressure. The NH_3-air mixture will combust very late compared to methane air mixture on introducing into combustion chamber due to very low flame speeds. Various studies on the laminar burning velocities are summarised in Table 5.2.

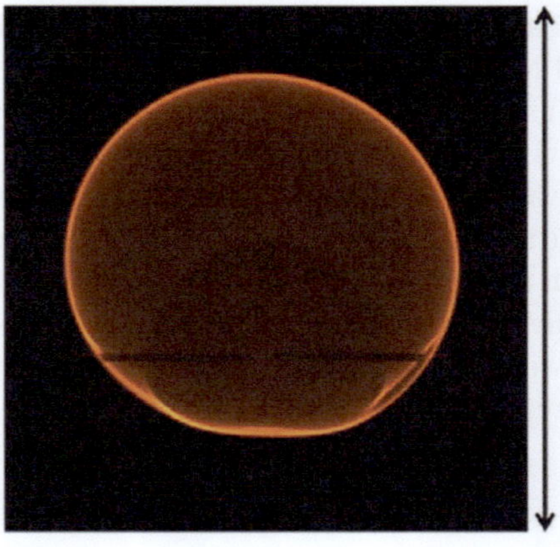

Fig. 5.1 Spherical NH_3-air laminar premixed flame (Kobayashi et al. 2019a)

The studies listed are both experimental as well as numerical. For experiments, a wide variety of methods are employed of which constant volume combustion chamber is a preferred choice. In numerical studies, different mechanisms are used to simulate the burning velocity and evaluate the effect of temperature, pressure and equivalence ratio. The most common mechanisms employed for simulation are Lindsted mechanism (1995), GRI 3.0 (Smith et al. 2024), Konnov mechanism (2009), Miller mechanism (1989) and lastly the Tian mechanism (2009). The first mechanism was developed by Miller in the year 1983 which involved 22 species and involved 98 reaction steps. Though it was a first attempt to predict the combustion phenomenon in ammonia, it significantly overpredicts the burning velocity (Hayakawa et al. 2015b). As the computational capability has progressed, the other mechanisms were developed by incorporating more species and reaction steps. A detailed comparison of all the mechanisms is described later in the chapter (Fig. 5.2).

It is clear from the experimental and analytical studies in Table 5.2 that the ammonia has a very low burning velocity, which is a challenge in adopting ammonia as a fuel in the current form. To overcome this problem, a few researchers have suggested methods like burning ammonia in oxygen rich environment (Wang et al. 2020). The increase in oxygen content in the ammonia-air mixture increases the reaction rates of radicals formed namely the OH, H, O, NH_2 and speeds up the process of burning (Shrestha et al. 2021). The studies on ammonia flame speed with the increased oxygen concentration is shown in Table 5.3. In general, for about 35–40% more oxygen than the stoichiometric requirement, the ammonia flame speeds can be matched with the methane-air flame speeds.

Another method to realise high flame speed with ammonia is to blend it with ammonia with high reactivity gaseous fuels methane or even hydrogen. For, ammonia-methane mixtures, the flame speed increases linearly with the concentration of methane (Shu et al. 2021). This is largely because of the increase in rate of reaction where H and OH radicals are formed. While for hydrogen-ammonia blending, the relation between flame speed and hydrogen concentration is exponential. At around 0.45 hydrogen concentration, the flame speed reaches comparable to methane air mixture, i.e. 35 cm/sec (Ichikawa et al. 2015; Han et al. 2019b).

Lastly, two important parameters critical for flame studies are the Lewis number and Markstein length. Lewis number is expressed as the ratio of thermal to mass diffusivity. It helps in characterising the dynamics and stability of the flame. The Lewis number for ammonia-air flame is < 1. Markstein length signifies the effect of curvature on the flame. Hayakawa et al. (2015c) measured the effect of equivalence ratio and pressure on Markstein length for ammonia flames and found that it increases with the increase in equivalence ratio like that of hydrogen flame. It is negative for equivalence ratio < 1. The burnt gas Markstein length increases with the pressure. However, it increases initially at the faster rate up to 0.3 MPa and then the rate of increase declines.

Table 5.2 Summary of flame studies on ammonia

References	Conditions	Methods	Findings
Cohen (1987)	T = 323–423 K	Flat flame burner	Increasing temperature increases flame speed by speeding the oxidation process
Zakaznov et al. (1978)	Ambient pressure & Room temperature	Cylindrical-tube method	Max flame speed observed around $\varphi = 1.1$
Ronney (1988)	Ambient & P = 0.066, 0.132, 0.329, 1 and 1.974 atm	Microgravity conditions in a constant volume combustion chamber (cylindrical)	Max flame speed of 8 cm/s observed. Slight overprediction for $\varphi > 1$ compared to experimental data
Pfahl et al. (2000)	Ambient pressure & Room temperature	constant volume combustion chamber equipped with schlieren imaging	Overprediction of flame speed at $\varphi < 1$ and underprediction at $\varphi > 1$
Takizawa et al. (2008)	Ambient pressure & Room temperature	Spherical constant volume combustion chamber	Flame speed in good agreement with experimental data
Duynslaegher et al. (2009)	P = 1–40 atm., T = 300K and $\varphi = 1.0$	Konnov mechanism	Increase of pressure decreases the flame speed
Hayakawa et al. (2015c)	Ambient pressure and Room temperature	Constant volume combustion vessel equipped with schlieren imaging	For $\varphi = 0.9$–1.2, Increase of pressure decreases the flame speed
Han et al. (2019a)	Ambient pressure & Room temperature	Heat flux method	Reduced kinetic model predicts flame speed with sufficient accuracy
Liu et al. (2019)	P = 0.3 - 1.6 atm., $\varphi = 0.5$–1.3, T = 298 K	Constant volume combustion vessel equipped with schlieren imaging	Increasing pressure speeds up the flame velocity marginally and then decreases continuously
Mei et al. (2019)	Ambient pressure & Room temperature	Constant volume combustion chamber (cylindrical)	Increasing the oxygen content increases the flame speed considerably
Li et al. (2014, 2015)	P = 1 atm., T = 298–573 K	Numerical analysis	Laminar burning velocities increase till $\varphi = 1.1$, and then decrease
Lhuillier et al. (2020)	P = 1 atm., $\varphi = 0.8$–1.4, T = 300–473 K with hydrogen doping	Spherical flame technique using schlieren imaging	Increasing the hydrogen concentration significantly increases the flame speed of ammonia

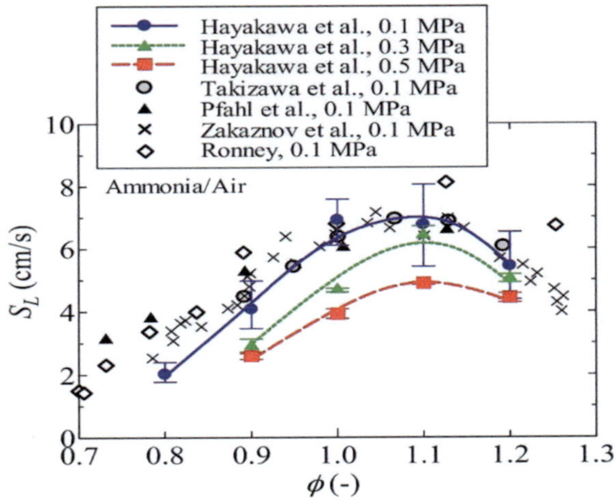

Fig. 5.2 Laminar flame speed (Kobayashi et al. 2019a)

Table 5.3 Ammonia combustion studies with increased oxygen concentration

References	Conditions	Methods	Results
Li et al. (2015)	O_2 conc. = 0.27–0.30	Numerical investigation using Miller mechanism (Miller and Bowman 1989) and the Reductive Konnov mechanisms (Duynslaegher et al. 2009, 2012)	Laminar is enhanced by 23% on increasing O_2 content by 11%
Takeishi et al. (2015)	O_2 conc. = 0.28–0.35	Experimental	Maximum velocity increases 500% (36.1 cm/s), compared to air only combustion
Mei et al. (2019)	O_2 conc. = 0.21–0.45	Constant pressure spherical flame method	The laminar burning velocity decreases with pressure and increases with the concentration of O_2
Wang et al. (2021a)	O_2 conc. = 0.6–1.0	Experimentation in constant volume vessel	For low O_2 concentration, The flame velocity is excessively dependent on temperature, and it increases with the temperature
Kim et al. (2021)	O_2 conc. = 0.35–0.40	Shadowgraphy in spherical stainless-steel chamber	Maximum velocity is 35.8 cm/s for given O_2 conc

5.2.2 *Ignition Delay*

Ignition delay (Ign. delay) is an important parameter of any fuel. Ignition delay period of ammonia have been investigated on variety of facilities like flow reactor, shock tubes and rapid compression machines. Numerically, Mathieu and Petersen (2015) proposed a mechanism consisting of 159 reaction steps and 35 species which was an extension of the work of Dagaut et al. (2005). The proposed mechanism was verified with experiments in the shock tube for temperatures 1550–2450K for both fuel rich and lean mixtures. They showed that for the autoignition of ammonia there is an interplay between the NH_2 and NO as follows:

$$NH_2 + NO \leftrightarrow NNH + H_2O\,[42] \tag{5.1}$$

Motivated by the above results, they extended their study to ammonia-hydrogen–oxygen and ammonia oxygen premixed flame at a relatively lower temperature (appx. 1000K). They concluded that apart from NH_2 & NO in Eq. (1), another reaction step which helps in precise prediction is as follows:

$$H_2NO + O_2 \leftrightarrow HNO + HO_2 \tag{5.2}$$

5.2.3 *Comparisons of Flame Characteristics Between Ammonia/Air and Methane/Air Mixtures*

Ammonia is generally seen as an alternative to the carbonaceous fuels to reduce the carbon emissions. While hydrogen is perceived as an ultimate fuel, the chemically closest carbonaceous compound to ammonia is methane. A comparison of ammonia, hydrogen and methane as a fuel is important to understand their combustion characteristics. Hydrogen is not feasible at present while methane is an established fuel worldwide. Thus, to shift from methane to non-carbon fuels like ammonia, the evaluation of its flame structure is essential.

Ammonia manifests a distinct combustion behaviour than methane and hydrogen respectively. Figure 3a, b shows the premixed hydrogen, ammonia and methane flames with air at stoichiometric concentration respectively. The flame structure was obtained through numerical simulation employing CEU-NH3-Mech 1.0 mechanism (Wang et al. 2021b) for ammonia, hydrogen and GRI Mech 3.0 (Smith et al. 2024) for methane-air mixtures for one-dimensional flame propagation. The data obtained was mole fraction, heat release and temperature at different locations of the flame. The prominent feature is the length of different flames. The heat release rate (HRR) here is an indicator of the amount of energy released by the fuel per unit volume and time. From heat release data it is evident that ammonia air flame has the least heat release rate, while hydrogen has the highest release. Methane heat release rate lies in between ammonia-air and hydrogen-air heat release rates. Similarly, for temperature, highest flame temperature is seen for hydrogen-air flame, while lowest flame temperature is for ammonia-air mixture. Though the difference is small, the reason for that is the lowest heating value of the ammonia in contrast to other two fuels. Moreover, ammonia-air generates a thick flame when compared to all other flames having a wide span of different species. The approximate length of active reaction zone of hydrogen is 0.04–0.1 cm. Also, the concentration of O, H, OH and NO radicals is very high in comparison to NO_2 and HO_2 radicals and hence their concentration greatly influence the reactivity of the flame. The decline in concentration of light H, O and OH pool radicals along with temperature near to the reaction zone supresses the flame reactivity and therefore the burning velocity also reduces. On the other hand, for methane air mixtures, as the methane is consumed, several intermediate species like CO are formed. This CO is oxidised to form CO_2.

In the pathway, the carbon containing species CH, CH_3, etc. are formed which are simultaneously liberated and absorbed by the flame in a narrow zone. For ammonia-air flame, the NO concentration is enhanced in the reaction zone and sharply declines at the downstream. The NO concentration is higher in ammonia-air flame compared to hydrogen-air and methane-air flame (Law 2006). At the downstream, the reactions proceed in the reverse direction due to fall in the concentration of O radical leading to decline in NO concentration. Looking this, it can be inferred that for ammonia-air flame, NO is dominantly produced from the fuel NO channel (Kobayashi et al. 2019b). Contrary to this, the NO formed in hydrogen-air and methane-air flame is mainly through Zeldovich mechanism ($N_2 + O \rightarrow NO + N$ and $N + O_2 \rightarrow NO + O$). The NO concentration in both mixtures increases moderately in the post flame region and declines downstream just like in the ammonia-air flame.

5.3 Chemical Kinetics of Ammonia Combustion

Chemical kinetics is an essential characteristic of fuel burning. It is described action mechanisms combining the transport coefficients and thermodynamics of the reacting species. The study on chemical reaction involved in flames of ammonia-air mixture started during 1960s (Fenimore and Jones 1961) at different conditions and with different compositions. The work of Miller and Bowman (1989) was the first detailed mechanism on the ammonia oxidation consisting of details of species and chemical processes involved in ammonia oxidation. The NOx formation was an important highlight of this mechanism. In Miller's mechanism there are 22 species participating in 98 sub-reactions. The first elementary reaction consisted of ammonia oxidation, followed by NO formation and reburn. This mechanism is in good agreement with the experimental data at lean and slightly rich conditions, for a wide range of equivalence ratio, pressure and temperature. The model successfully predicts the experimental works by MacLean and Wagner (1967). Since then, several reaction models have been proposed. The other reaction models developed are summarised in Table 5.4.

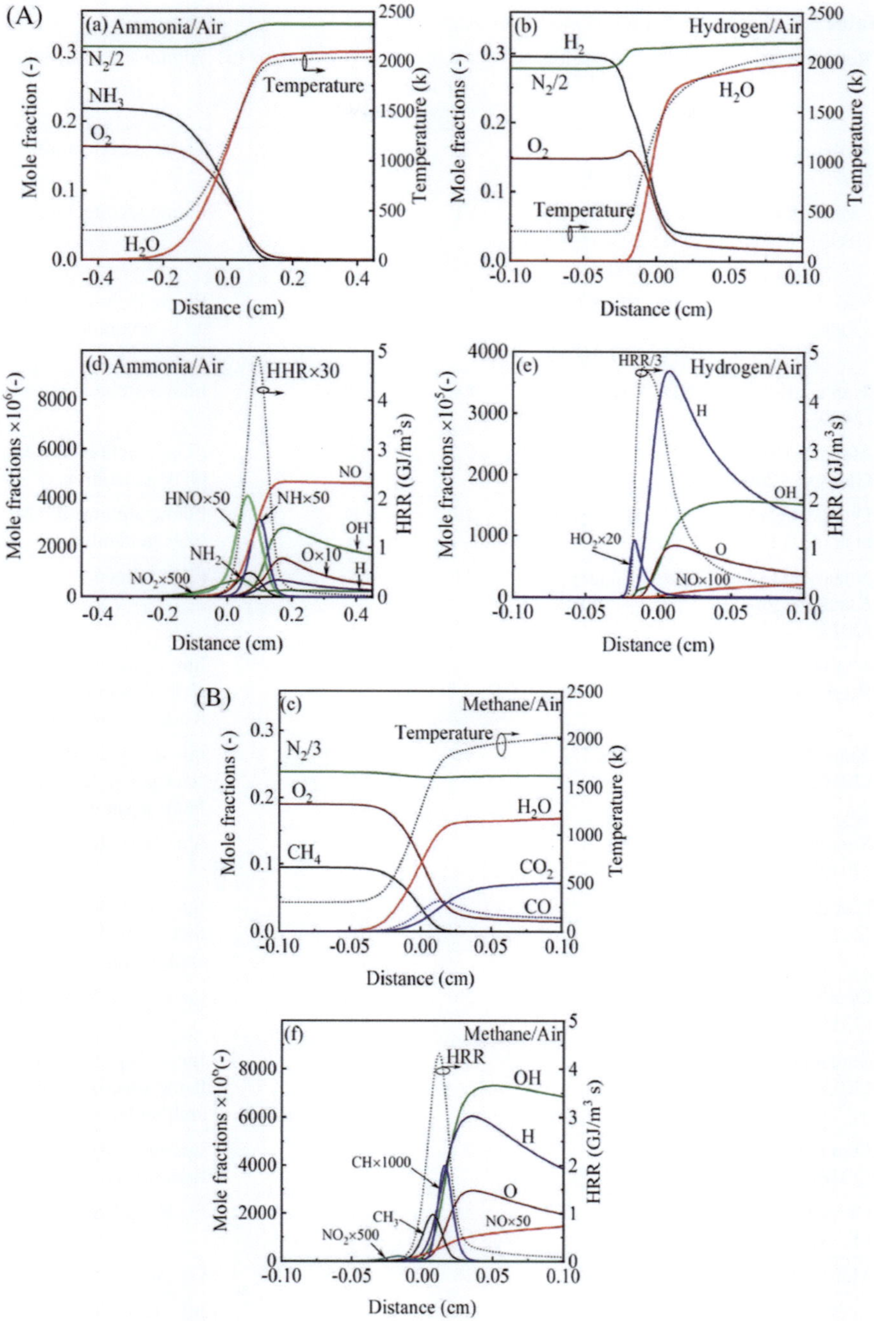

Fig. 5.3 **a** Premixed flame structures of ammonia-air, hydrogen-air at stoichiometric mixtures and 1 bar pressure **b** Premixed flame structures of methane-air mixture at stoichiometric mixtures and 1 bar pressure (Elbaz et al. 2022)

Table 5.4 Summary of reaction models of ammonia

References	Mixture constituents	No. of reacting species	No. of steps in reaction	Parameters evaluated
Miller et al. (1983)	[NH_3, O_2], [NH_3, H_2, O_2]	22	98	Flame architecture
Lindstedt et al. (1995)	[NH_3, H_2, O_2, N_2], [NH_3, NO, H_2-O_2], [NH_3, O_2]	21	95	Flame architecture
Skreiberg et al. (2004)	[NH_3, NO, CO, H_2, O_2, N_2], [NH_3, CO, H_2, O_2, N_2]	34	191	Flame architecture & NO_x generation
Tian et al. (2009)	[NH_3, CH_4, O_2, Ar],	84	703	Flame architecture
Mendiara and Glarbory (2009)	[NH_3, CH_4, O_2, CO_2, N_2]	97	779	Flame architecture & NOx generation
Duynslaegher et al. (2012)	[NH_3, H_2, O_2, Ar]	19	80	Flame architecture & NOx generation
Nozari and Karabeyoğlu (2015)	[NH_3, H_2, air]	21	91	Flame speed
Mathieu and Petersen (2015)	[NH_3, O_2, Ar]	35	159	Ign. delay duration, NOx generation, & flame architecture
Xiao et al. (2016)	[NH_3, CH_4, O_2, Ar]	48	500	Ign. delay duration, NOx generation, & flame architecture
Song et al. (2016)	[NH_3, O_2]	32	204	Ign. delay duration
Nakamura et al. (2017)	[NH_3, air]	38	232	Ign. delay duration, flame speed & flame architecture
Okafor et al. (2018)	[CH_4, NH_3, air]	59	365	Laminar flame speed
Shrestha et al. (2018)	[NH_3, air]	34	264	Ign. delay duration, flame speed, & flame architecture
Otomo et al. (2018)	[NH_3, air], [NH_3, H_2, air]	32	213	Ign. delay duration & flame speed
Okafor et al. (2019)	[CH_4, NH_3, air]	38	140	Flame speed
Mei et al. (2019)	[NH_3, O_2, N_2]	38	265	Flame speed
Li et al. (2019)	[NH_3, H_2, air], [NH_3, CH_4, air]	28 51	213 420	Ign. delay duration, flame speed & flame architecture

(continued)

Table 5.4 (continued)

References	Mixture constituents	No. of reacting species	No. of steps in reaction	Parameters evaluated
Stagni et al. (2020)	[NH_3, O_2, Ar] [NH_3, air]	31	203	Flame speed velocity & flame architecture
Han et al. (2020)	[NH_3, syngas, air]	35	177	Flame speed & flame architecture
Wang et al. (2022)	[NH_3, CH_3OH, air], [NH_3, C_2H_5OH, air]	91	444	Flame speed
Singh et al. (2021)	[NH_3, air]	32	259	Ign. delay duration, flame speed, & flame architecture
Issayev et al. (2022)	[NH_3, DME, air]	176	1419	Ign. delay duration & flame speed
Shrestha et al. (2022)	[NH_3, DEE, air]	280	2238	Flame speed, Ign. delay duration & flame architecture
Gotama et al. (2022)	[NH_3, H_2, air]	26	119	Flame speed
Cai et al. (2022)	[NH_3, air] [NH_3, H_2, air]	20	89	Flame speed & Ign. delay duration
Yang et al. (2023)	[NH_3, CH_4, air]	187	1571	Soot volume fraction

It is clear from Table 5.4, that in the initial years, the accuracy of mechanisms was evaluated from the flame structure. But, with the advent of modern techniques for measurement, the mechanisms are being assessed by parameters like ignition delay time, laminar burning velocity, etc. Certain landmark studies have also evaluated the NOx formation in ammonia owing to inherent nitrogen in the ammonia fuel. This has been a growing concern of the environmentalists and a major hindrance in successful deployment of ammonia as an alternative fuel for IC engines. The NOx formation and mitigation strategies used for ammonia combustion are discussed in the subsequent sections.

5.3.1 Ammonia Decomposition Pathway

Ammonia when combusted lean, there is a massive pool of OH and O radicals which promote NO formation from NH_y. Ammonia decomposition generally happens primarily with H abstraction. The replacement of one hydrogen in NH_3 is called H abstraction. The H abstraction reactions are generally endothermic in nature. At temperatures less than 400K, such reactions can be easily ignored but at high temperatures they have a sizable rate constant. The principal pathway for ammonia decomposition through H radical is as per Eq. (5.3).

$$NH_3 + H \rightarrow NH_2 + H_2 \tag{5.3}$$

These reactions are confirmed by several studies, but the difference lies in rates of these reaction. The most well-accepted mechanism for Eq. (5.3) is from Michael et al. [69A] at low temperatures while for higher temperatures, Glarborg-Mech (2021) is found to yield better results. Moreover, Stagni et al. (2020) determined the rate constant but it was found to be twice as compared to Michael et al. (1986) in the temperature range 350K-2500K. Additionally, Shrestha-Mech (2018) predicted the rate constants to be one-third to that of Stagni et al. (2020).

Another important H abstraction happens through the OH radical as described by Eq. (5.4).

$$NH_3 + OH \rightarrow NH_2 + H_2O \tag{5.4}$$

Here also there is a significant difference of rate constant reported in the literature. The rate constant calculated from CEU-NH3 was approximately half of the findings of Stagni et al. (2020). Also, Shrestha-Mech (2018) and Glarborg-Mech predicted the rate constant of Eq. (5.4) to be 1.25 times as reported by CEU-NH3. Samu et al. (2018) proposed that, therefore, an appropriate mechanism should be selected for accurate predictions.

The second sub reactions of interest are represented by following equations

$$NH_2 + H \rightarrow NH + H_2 \tag{5.5}$$

$$NH_2 + OH \rightarrow NH + H_2O \tag{5.6}$$

These reactions predominantly occur at rich conditions. Here also, there is lack of concurrency in the rate constant and care must be taken to select the appropriate mechanism especially at high temperatures.

5.3.2 *Ammonia Oxidation Pathway*

Since ammonia has an inherent nitrogen molecule, the ammonia oxidation reactions are also termed as the NOx formation reactions. After the H abstraction as discussed in the above section, the radicals interact amongst themselves. The NOx formation is primarily due to interaction of N, NH, NH_2 with HO_2, OH and OH radicals. In this process, certain intermediate species (H_2NO, HNO) are formed which is shown in Fig. 5.4. The major reactions with the NH_2 radical here being

$$NH_2 + O \rightarrow H_2NO \tag{5.7}$$

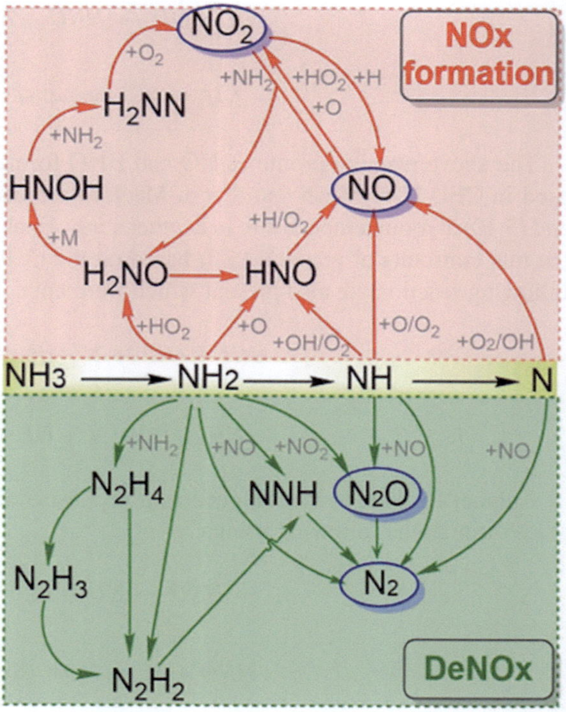

Fig. 5.4 NOx and DeNOx pathways on combustion of ammonia (Elbaz et al. 2022)

$$NH_2 + O \rightarrow HNO + H \tag{5.8}$$

$$H + O_2 \rightarrow O + OH \tag{5.9}$$

These above reactions are less temperature and pressure sensitive and have high sensitivity to the concentration of NO. However experimentally they have been only verified at low temperatures. The flame speed is highly sensitive to (5.8) especially when H_2 is doped in ammonia as described in the works of Shrestha et al. (2018). More experimental data are needed to have a clarity of the process at high temperatures. In general, Stagni-Mech and CEU-NH3-Mech are widely used to predict the oxidation capability of ammonia (Sumathi et al. 1998).

After NH_2 radical mechanisms, comes the NH radical mechanisms. These can be majorly summarised as follows:

$$NH + O_{2-} \rightarrow HNO + O \tag{5.10}$$

$$NH + O_2 \rightarrow NO + OH \tag{5.11}$$

$$NH + OH \rightarrow HNO + H \tag{5.12}$$

$$NH + O \rightarrow NO + H \tag{5.13}$$

The above reaction promotes NO and HNO formation. The rate constants were used in CEU-NH3-Mech and Stagni-Mech to evaluate rate constants for the (5.9) and (5.10) at room temperature. In another study, Elishav et al. (2020) reappropriated the rate constants of reaction (5.9) based on the (5.10) and (5.11) and showed the following reaction are also present which were not considered earlier.

$$NH + OH \rightarrow N + H_2O \tag{5.14}$$

$$NH + O \rightarrow N + OH \tag{5.15}$$

Another oxidation reaction occurring is conversion of HNO to NO which takes place through the following means

$$HNO \rightarrow NO + H \tag{5.16}$$

$$HNO + H \rightarrow NO + H_2 \tag{5.17}$$

$$HNO + OH \rightarrow NO + H_2O \tag{5.18}$$

The reaction rates of above reaction are highly dependent on the mode of reaction. Researchers like Stagni et al. (2022) have assumed it to be direct decomposition reaction while others (Shrestha-Mech and Glarborg-Mech) have considered to be bimolecular or a three-body collision reaction. The reaction rates of unimolecular decomposition are assumed to be lower than bimolecular or three body collisions. Like NOx formation, a wide range of opposite reaction also takes place which convert the NOx formed into N_2, which is an inert gas. These reactions are also important to study to deNOx the ammonia combustion process. Such reactions will be discussed in detail under subsequent section in the chapter relating to emission reduction strategies.

5.3.3 NH_y Recombination Pathway

Under the combustion conditions, the ammonia decomposes to form a highly reactive NH_y radicals. Such radicals, under fuel rich conditions, participate in the reduction reaction to produce N_2. The pathway followed is generally N_2H_y route as shown in the following reaction.

$$NH_2 + NH_2 \rightarrow N_2H_4 + H \tag{5.19}$$

$$NH_2 + NH \rightarrow N_2H_2 + H \tag{5.20}$$

$$NH_2 + NH \rightarrow N_2H_3 \tag{5.21}$$

$$NH_2 + NH_2 \rightarrow N_2H_2 + H_2 \tag{5.22}$$

NO production lower in rich flames because the concentration of O and H radicals is lower in rich flames, and these N_2H_y reaction abate the NO formation. In addition to this, Klippenstein et al. (2009) advocated the following reaction channel.

$$NH_2 + NH_2 \rightarrow NH_3 + NH \tag{5.23}$$

Reaction (5.19) is favoured at temperature below 600K and pressure above 1 bar and reaction rates given by Klippenstein et al. (2009) are good. Elishav et al. (2020) gave the reaction rates with good accuracy for high temperature conditions. Majority of reactions of N_2H_y in Shrestha-Mech are adopted from Dean and Bozzelli (2000) while CEU-NH3-Mech used inputs from Glarborg-Mech and Stagni-Mech.

Reactions (5.20) and (5.21) are sensitive to equivalence ratio greater than 1. The reaction (5.20) is certain but in very few studies, it is also perceived to have an alternate pathway. One is shown in (5.21) and in another it produces N and NH_3 as products. Elishav et al. (2020) compared the rate coefficients proposed by Davidson et al. (1990) with the Klippenstein et al. (2009) and concluded that coefficients given by Davidson et al. (1990) agreed well for the experimental data.

The sensitivity of laminar burning velocity for rich and lean mixtures as indicated by CEU-NH3-Mech in the mixture of NH_3, syngas and air. The (5.19) is a branch initiation reaction while (5.20) terminates the reaction. (5.20) is preferred at rich flames. The reaction (5.20) is only included in CEU-NH3-Mech mechanism. Consequently, CEU-NH3-Mech shows better predictability of experiments involving rich flames. Some other reactions involving NH_2 are summarised in the following Table 5.5.

Another radical participating in the ammonia combustion process is the NNH radical. NNH radical can undergo decomposition reactions through two routes. It

Table 5.5 Summary of other NH_2 reactions

Reaction	Original source	Including mechanisms
$NH_2 + NH_2 \rightarrow N_2H_3 + H$	Dean and Bozzelli (2000)	Stagni-Mech, Shrestha-Mech and CUE-NH3-Mech
$NH_2 + NH_2 \rightarrow N_2H_2 + H_2$	Klippenstein et al. (2009)	Stagni-Mech, Shrestha-Mech and CUE-NH3-Mech
$NH_2 + NH_2 \rightarrow H_2NN + H_2$	Klippenstein et al. (2009)	CUE-NH3-Mech

can either undergo unimolecular decomposition to generate N_2 and H or can undergo bimolecular decomposition to produce O_2 and H.

$$NNH \rightarrow N_2 + H \tag{5.24}$$

$$NNH + O_2 \rightarrow N_2 + HO_2 \tag{5.25}$$

The NNH decomposition is also crucial for the thermal deNOx process in which NH_3 is used for selective non catalytic reduction of NO. The efficiency of this process is largely dependent on the lifespan of the NNH radical and temperature at which the reaction occurs. Thus, this mechanism becomes important to consider during short time spans and lower temperatures.

5.3.4 Effect of Temperature and Pressure on Ammonia Combustion

Effect of Temperature

Temperature less than 1000K are termed as low temperatures for combustion. The low temperature effects the ignition delay of the fuel. At low temperatures, the reactions involving H_2NO and HNO play a crucial rule. H_2NO along with HNO mechanisms are a must in ignition delay time prediction of ammonia. At low temperatures H_2NO is generated through the reaction of NH_2 with HO_2 and NO_2 as follows.

$$NH_2 + HO_2 \rightarrow H_2NO + OH \tag{5.26}$$

$$NH_2 + NO_2 \rightarrow H_2NO + NO \tag{5.27}$$

Several simultaneous chain branching and propagating reactions also take place which consume H_2NO and HNO as follows:

$$H_2NO + O_2 \rightarrow HNO + HO_2 \tag{5.28}$$

$$HNO + O_2 \rightarrow NO + HO_2 \tag{5.29}$$

$$NO + HO_2 \rightarrow NO_2 + OH \tag{5.30}$$

At higher temperatures, the chemistry changes. The reaction $NH_2 + NO_2$ dominates over $NH_2 + NO$ reaction. The conversion of NO to NO_2 is supressed due to lesser amount of HO_2 radicals. The verification of this was obtained by the results of ignition delay duration measurement by He et al. (2019) within the temperature range 1080–1437K. These results are well incorporated in Glarborg-Mech mechanism which predicts well in the same temperature range. On further increasing the temperature, new reactions like decomposition (of NNH, NH and N_2O) start taking over which increase the overall reactivity.

Effect of Pressure

The NH_y recombination reaction responds to pressure variations but little has been studied about their pressure sensitivity. However, some literature is available on the following reaction

$$NH_2 + H \rightarrow NH_3 \tag{5.31}$$

Stagni et al. (2020) studied the above recombination reaction initially and the work was extended by Glarborg et al. (2021) incorporating the three body impact efficiencies of the above and N_2H_4 recombination reaction. Dai et al. (2021) also studied the ignition delay with respect to N_2H_4 recombination reaction and concluded that it can be neglected with respect to different pressures.

In case of NOx formation, increasing pressure reduces the OH and O radicals which in turn reduces the NO radicals (Okafor et al. 2018). Hayakawa et al. (2015c) have showed that increasing pressure suppresses the rate of H, OH conversion into H_2O and hence the chances of NOx production is minimised. Moreover, increasing pressure increases the pathway of NH_2 radicals taking the N_2 route and large number of HO_2 radicals. These radicals participate less in the NOx formation due to less reactivity.

5.4 Emissions Reduction in Ammonia Combustion

The emission during ammonia combustion is largely NOx. Other emissions like PM, unburnt hydrocarbons do not directly emanate from ammonia but other fuels which are blended to enhance the combustion properties of ammonia.

5.4.1 DeNOx Strategies in Ammonia Combustion

NOx poses a serious risk to the human health. Growing environmental concerns have led to an increase in stringent emission norms which limit the NOx. The NOx aftertreatment systems are complex and increase the overall cost of the system considerably. Thus, it is imperative to reduce the NOx right from the start of combustion and fine tune the combustion parameters in a way that NOx is minimised. Some studies are available which propose modifications like altering the combustion chamber design (Medhat et al. 2021), composition of fuel (Brackmann et al. 2016), fuel mixing

(Gotama et al. 2022) and staged combustion (Medhat et al. 2021). These fundamental studies generally employ the use of numerical means (DNS) or experimental means (PLIF-planar laser-induced fluorescence). Such strategies are discussed in detail in the forthcoming sections.

Dual Fuel Combustion

The dual fuel combustion is one of the effective ways to utilise the ammonia in a combustion system. It not only overcomes the poor combustion properties of ammonia, but also seems effective in minimising the NOx from the ammonia-based combustion. For the latter, the fraction of ammonia substituting the reactive fuel plays a critical role in emissions. For example, in a study by Rocha et al. (2019), the NOx emissions from two mixture fuel compositions were discussed. One was the mixture of ammonia, methane and air while the other was ammonia, hydrogen and air. The NOx emission increases initially by increasing the content of ammonia and later declines. The NOx emissions peak at 50% ammonia concentration due to dependence of NO formation on the methane oxidation process. The CH_4 consumes the oxygen rich radicals during oxidation. On the contrary, with mixture containing ammonia, hydrogen and air, the reaction involving hydrogen oxidations is overpowered by oxidation of NH_2 and NH radicals. Consequently, lower NOx is observed with hydrogen mixture at same concentration of ammonia. In general, the NOx formation is guided by the race of chemical and dilution effects. The kinetic reaction mechanisms currently available consider these effects but generally overpredict the effect (Xiao et al. 2016). Thus, more mechanisms with validated with substantial experimental data need to be developed.

Fuel-Rich Combustion

Rich combustion is also one of the promising solutions in abatement of NOx. Though it increases the fuel cost but is still cheaper in comparison to employing costly aftertreatment devices. The rich combustion (till $\varphi = 1.1$) minimises the highly reactive O and OH radicals and therefore minimises the chances of N taking the NO route (Ni et al. 2021). The main drawback of this method is the chances of unburnt ammonia in the emissions or ammonia leakage through combustion chamber. Since ammonia is a highly toxic fuel, use of such strategy poses a serious risk.

Use of High Inlet Pressures

Another method of quelling the NOx emission is by increasing the inlet pressure. Studies using both numerical (Cai et al. 2022) and experimental PLIF technique (Kobayashi et al. 2019b) are available to get a better understanding of the chemistry at high inlet pressures up to 5 bar. The NH_2 radical during ammonia combustion is largely responsible for NO. At high inlet pressures, the NOx reduces due to lower concentration of NH and NO radicals. The NO generating reaction are not favoured at high pressure conditions. Also, the formation of OH radicals are supressed hindering the following reaction

$$HNO + OH \rightarrow NO + H_2O \tag{5.32}$$

In place of above sub reaction, the H and OH radicals combine to form water at high pressures resulting in a cleaner combustion. Increasing inlet pressure is an effective method to reduce NOx, but excessive inlet pressure can further reduce the laminar burning velocity. This can make the combustion unstable. Thus, care must be taken to select adequate inlet pressure to compromise between NOx emission and flame speed.

Water Addition

Another common method to decrease NOx emission is addition of water vapour during combustion. The water vapour reduces the flame temperature which in turn reduce the NOx emission. Experiments have shown that introduction of water in the combustion chamber can accelerate the dissociation of H_2O with O, increasing the pool of OH radicals. As a result, less O radicals are available for N to react and form NO (Law 2006). Ariemma et al. (2021) performed experiments to measure the effect of humidification on NOx and combustion temperature. The combustion temperature continuously decreased with increasing the water content in air (relative humidity) from 0 to 75%. In addition to this, the effect of equivalence ratio was studied on water addition. The water addition was beneficial only in fuel lean conditions because under lean conditions, the O radicals are present in large numbers which react with H_2O, and NO formation is minimised. In contrast to this, water addition has little effect on rich combustion. Moreover, during slightly rich combustion, non-premixed combustion leads to higher NOx emissions than premixed whereas opposite is seen during lean and stoichiometric combustion. Such results are helpful in designing suitable injection scheme for ammonia combustion.

5.4.1.1 Staged Combustion

Like in other fuels, staged combustion of ammonia also leads to lower NOx output. In staged combustion, the fuel is introduced in the chamber in a way such that it creates two separate zones. One is a fuel rich region while the other is fuel lean region (Zhang et al. 2022), creating two different equivalence ratio zones. Experiments like that of Okafor et al. (2019) have shown that NO emissions are controlled mainly by primary equivalence ratio. The NOx is minimised in ammonia-methane-air mixture when the primary equivalence ratio is approximately 1.3, i.e. mole fraction of ammonia is 0.1. This is because of non-abundancy of OH and O radicals due to low concentration of ammonia. Combining elevated inlet pressures and rich primary equivalence ratio can lead to very high combustion efficiency and ultra-low NOx.

5.4.2 Soot Abatement in Ammonia Combustion

Soot is generally formed by partial combustion of carbonaceous fuels. Combustion of pure ammonia does not emit soot. Soot is produced when ammonia is blended with carbon-based fuels. In hydrocarbons, soot evolution starts in two steps. At first, the pyrolysis of precursors occurs, resulting in intermediate species (commonly acetylene or ethylene) and in the second step polycyclic aromatic hydrocarbons (benzene) are created. Thus, to assess soot evaluation, the above compounds (acetylene/ethylene/benzene) are considered as reference in initiating soot formation.

The introduction of ammonia along with other hydrocarbon fuel changes the chemistry of soot formation and needs investigation (Yang et al. 2023). The earliest works on the effect of ammonia on hydrocarbon fuel was done by Bockhorn et al. (1981) by use of propane. Addition of maximum 3.5% ammonia to propane resulted in lower soot and the size of soot particles was greatly reduced. For liquid fuels, the studies reported in literature are largely performed with n-heptane, which is a diesel surrogate fuel. The reason being diesel is the preferred fuel for dual fuel engine due to its high reactivity (Agarwal et al. 2022). Gülder et al. (1996) analysed the soot formation by cofiring the n-heptane diffusion flames doped with ammonia in a burner in argon environment. The 2D laser induced fluorescence method was used to quantify the OH and polycyclic aromatic hydrocarbons while the soot volume fraction was measured using laser induced incandescence approach. The polycyclic aromatic hydrocarbons and soot were found to be surrounded by the OH radicals. The frequency of OH radicals was reducing in the streamwise direction. The minimum and maximum amount of soot was at the flame root and flame tip respectively while it was vice versa for polycyclic aromatic hydrocarbons. The concentration of soot, polycyclic aromatic hydrocarbons and OH radicals decreased on increasing the concentration of ammonia. The soot concentration decreased on 1:1 basis with the ammonia concentration due to inhibition effect on soot during inception phase of soot instead of oxidation. A similar trend was reported in the study, which compared the polycyclic aromatic hydrocarbons ratio in combustion of pure n-heptane and mixtures of n-heptane and ammonia. The noteworthy feature of this research was the study of interaction between polycyclic aromatic hydrocarbons and nitrogen which leads to soot suppression.

Another detailed numerical and experimental study was performed by Montgomery et al. (2021) which substituted ammonia in methane. A comparison was made between addition of nitrogen and methane to study the effect on soot volume fraction. Numerically it was observed that ammonia provided lower soot than nitrogen. Also, the soot volume fraction reduced to one-tenth with ammonia when 20% of ammonia was replaced by nitrogen on molar basis. The experiments were also performed using the colour ratio pyrometry. Similar results were obtained in addition to more orange flame at higher ammonia concentration due to NH_2 chemiluminescence. To have a better assessment of the effect of ammonia and nitrogen doping, the flame was compared to acetylene and benzene flames since these two compounds are directly responsible for soot formation. The fundamental parameter studied was height above burner. Generally larger flame height signifies late reaction rate. Addition of ammonia or nitrogen increased the flame height for all the fuels signifying lower reaction rates. Moreover, ammonia showed most hindrance to soot formation than acetylene and the distinctive feature with ammonia combustion is the decreased polycyclic aromatic hydrocarbons formation. The main limitations of the study were poor predictability in regions of soot formation and intermediate radicals in sub-reactions.

Wide range of studies on soot suppression by ammonia addition has also been reported with acetylene fuel. Renard et al. (2009) performed experiments with ethylene, oxygen, argon flames with ammonia substitution under rich conditions. The study demonstrated that the presence of ammonia has a stronger influence in development of C5-C10 in situ products than CO_2 and water. Zhou et el. (2022) also performed the study using planar laser-induced fluorescence and incandescence with ethylene and ammonia to evaluate the mechanisms of soot formation and the average particle diameter (D_{63}). In comparison to ethylene, the peak soot volume fraction decreased by 86.7% with just 0.2 mol fraction of ammonia. In addition to this, the average particle diameter drastically reduces with increasing the concentration of ammonia. When any hydrocarbon fuel is replaced with ammonia, three mechanisms act together to abate the soot formation, i.e. chemical effect, dilution effect and the thermal effect. The former being the most dominant in case of acetylene-ammonia mixture while the other two are weaker. Since the peak temperatures and temperature profiles in the above study were same with ammonia, the role of thermal effect was negated. The major influence was concluded to be of the chemical effect, i.e. the C-N chemistry. The minor role is played by the dilution effect due to lower availability of C atoms. It is also worth noting that these effects are also applicable during combustion of liquid fuels, but the degree is much more pronounced in gaseous fuels. Thus, it can be concluded that ammonia is more effective in soot suppression in gaseous fuels than liquid fuels.

In another study (Boyette et al. 2021) of the mixture of acetylene-ammonia and acetylene-hydrogen, turbulent diffusion flames were studied for soot formation. Laser-induced incandescence method was employed to assess the soot volume fraction and the UV bands along with pointwise laser-induced fluorescence were used to determine the soot precursors. The results showed that the soot distribution is originally Gaussian shaped at the centre of the flame and undergoes substantial transformation (shown in Fig. 5.5). The soot distribution can broadly distribute into three different regions namely soot inception region, rapid growth region and the oxidising region. In the first region, the soot concentration is nominal. In the next region there is a manifold increase in soot concentration because of surface expansion and particle clustering. In the last stage, the soot concentration is again low due to oxidation of soot particles due to mixing. Another aspect of interest is the soot particle size. The soot particles are initially smaller and gain diameter in the direction away from combustion nozzle up to a point and again shrink further on going in the same direction.

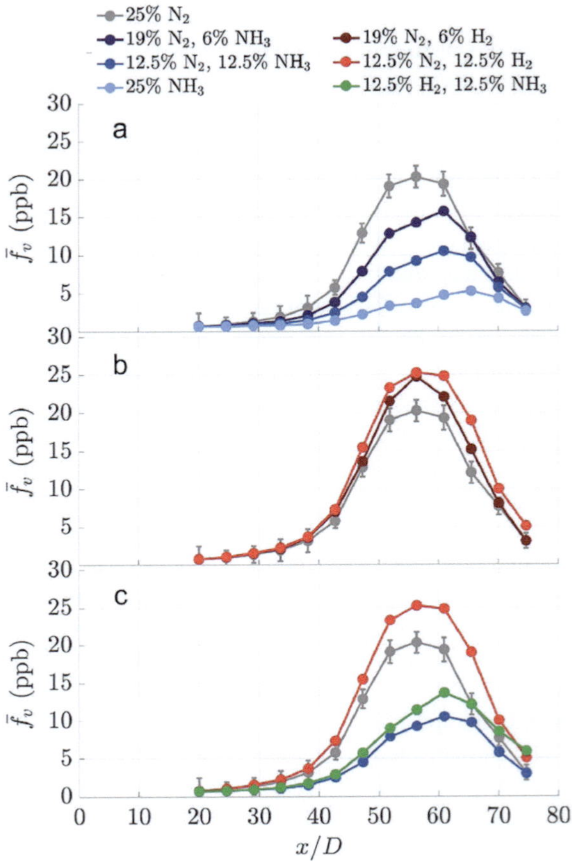

Fig. 5.5 Mean soot volume fraction of N_2-NH_3 and N_2-H_2 mixtures with flame location on the X-axis and soot volume fraction on Y-axis (Boyette et al. 2021)

5.5 Challenges in Ammonia Combustion

The utilisation of ammonia for transport sector is a crucial step in decarbonizing the economy (Zincir et al. 2023). Though a lot of scope exists but ammonia fuelled systems are not widely available due to various challenges. Ammonia is a highly toxic chemical and even slightest contact with the humans can cause severe injury and death. Thus, a robust handling system is needed for widespread use of ammonia as a transportation fuel. Next biggest challenge with ammonia is its physical and chemical properties. Properties like low calorific value, toxicity, lower flame speed, high ignition delay period, narrow flammability limit, low reactivity, etc. do not make ammonia a combustion ready fuel in the current system designed for fuels like gasoline and diesel. The ammonia combustion systems need to be tailor made, keeping in mind its unique properties. For this, the combustion chemistry needs to be investigated in detail. The available combustion mechanisms are not able to fully manifest the sub reactions of unstable radicals leading to inaccurate combustion analysis either by over predicting or under predicting some species. Another challenge is the changing combustion chemistry of ammonia which changes under rich and lean conditions. In rich conditions, NH_y and H radicals are abundant producing NNH because of NH_y recombination reactions. NOx formation can be minimised due to N_2 formation through N_2O and NNH route. Contrary to this, under lean conditions, O and OH radicals are abundant leading to NOx production through HNO route. These reactions are captured in the chemical mechanisms, but a high degree of uncertainty in rate constant of such mechanisms still prevails. Thus, there is a need of robust chemical kinetic models backed by experimental data.

5.6 Future Scope in Ammonia Combustion

Ammonia is an important carbon free fuel. Currently it is the only non-carbon-based fuel that offers significant advantages like high energy density, easy storage and transportation. Its production has matured; leading to sustainable supply systems worldwide. Therefore, it has a huge potential to be incorporated as an alternative to conventional fuels. The full potential of ammonia should be realised by studying the detailed chemical mechanisms. The mechanisms should incorporate the participation of various species to fully understand ammonia combustion. The mechanisms must be developed based on comprehensive direct numerical simulations rather than on reduced reaction model. The new mechanisms developed should be validated against a wide range of experimental statistics in the literature, for robust prediction of combustion parameters. Moreover, the rate constant of such important reaction should be evaluated to consistent prediction of the model.

Currently major problems which are associated with ammonia are high NOx emission, unburnt ammonia emissions and low flame speed. Such problems can be addressed by a variety of means discussed in this chapter. The thermal DeNOx strategy or rich combustion can be used to decrease the amount of NOx from combustion emissions. Use of modern experimental methods can be devised to propose accurate chemical mechanisms leading to better understanding of the overall combustion process. With this knowledge, the adoption of different DeNOx pathways instead of conventional ammonia combustion will lead to lower emissions. Additionally, when used in dual fuel configuration, the CO and CO_2 formation is also not understood fully. Thus, studies pertaining to the interaction of C-N, H_2NO, HNO, etc. during combustion will lead to better understanding of such emissions. It has already been proven that the flame speed of ammonia reaches the same level as methane when it is burnt in an oxygen rich environment. Therefore, such challenges can be easily tackled by innovative hardware engineering in combustion systems. The flame speed of ammonia can also be increased by doping it with hydrogen. But still, the formation rate of intermediaries like H_2NO is not readily available. Thus, the development of reliable mechanisms with high predictability is necessary for realising ammonia as a cleaner fuel for combustion.

References

Agarwal AK (2007) Biofuels (alcohols and biodiesel) applications as fuels for internal combustion engines. Prog Energy Combust Sci 33(3):233–271

Agarwal AK, Gupta T, Kothari A (2011) Particulate emissions from biodiesel vs diesel fuelled compression ignition engine. Renew Sustain Energy Rev 15(6):3278–3300

Agarwal AK, Singh AP, Maurya RK (2017) Evolution, challenges and path forward for low temperature combustion engines. Prog Energy Combust Sci 61:1–56

Agarwal AK, Kumar V, Jena A, Kalwar A (2022) Fuel injection strategy optimisation and experimental performance and emissions evaluation of diesel displacement by port fuel injected methanol in a retrofitted mid-size genset engine prototype. Energy 248(1):123593

Ariemma G, Sabia P, Sorrentino G, Bozza P, De Joannon M, Ragucci R (2021) Influence of water addition on MILD ammonia combustion performances and emissions. Proc Combust Inst 38(4):5147–5154

Bockhorn H, Fetting F, Meyer U, Reck R, Wannemacher G (1981) Measurement of the soot concentration and soot particle sizes in propane oxygen flames. Symp (Int) Combust 1137–1147

Boyette WR, Steinmetz SA, Guiberti TF, Dunn MJ, Roberts W, Masri AR (2021) Soot formation in turbulent flames of ethylene/hydrogen/ammonia. Combust Flame 226:315–324

Brackmann C, Alekseev VA, Zhou B, Nordstr¨om E, Bengtsson PE, Li Z, Ald´en M, Konnov AA (2016) Structure of premixed ammonia+ air flames at atmospheric pressure: laser diagnostics and kinetic modelling. Combust Flame 163:370–381

Cai T, Zhao D, Chan SH, Shahsavari M (2022) Tailoring reduced mechanisms for predicting flame propagation and ignition characteristics in ammonia and ammonia/hydrogen mixtures. Energy 260:125090

Cohen N (1987) The O+ NH_3 reaction: a review. Int J Chem Kinet 19(4):319–362

Dagaut P, Nicolle A (2005) Experimental and kinetic modeling study of the effect of SO_2 on the reduction of NO by ammonia. Proc Combust Inst 30(1):1211–1218

Dai L, Hashemi H, Glarborg P, Gersen S, Marshall P, Mokhov A et al (2021) Ignition delay times of NH3/DME blends at high pressure and low DME fraction: RCM experiments and simulations. Combust Flame 227:120–34. https://doi.org/10.1016/j.combustflame.2020.12.048

Davidson D, Kohse-H¨oinghaus K, Chang A, Hanson RK (1990) A pyrolysis mechanism for ammonia. Int. J Chem Kinet 22:513–35. https://doi.org/10.1002/kin.550220508

Dean AM, Bozzelli JW (2000) Combustion chemistry of nitrogen. Gas-Phase Combust Chem 125–341

Duynslaegher C, Jeanmart H, Vandooren J (2009) Flame structure studies of premixed ammonia/hydrogen/oxygen/argon flames: experimental and numerical investigation. Proc Combust Inst 32(1):1277–1284. https://doi.org/10.1016/j.proci.2008.06.036

Duynslaegher C, Contino F, Vandooren J, Jeanmart H (2012) Modelling ammonia combustion at low pressure. Combust Flame 159:2799–2805

Elbaz AM, Wang S, Guiberti TF, Roberts WL (2022) Review on the recent advances on ammonia combustion from the fundamentals to the applications. Fuel Commun 10:100053

Elishav O, Mosevitzky Lis B, Miller EM, Arent DJ, Valera-Medina A, Grinberg Dana A et al (2020) Progress and prospective of nitrogen-based alternative fuels. Chem Rev 120(12):5352–5436. https://doi.org/10.1021/acs.chemrev.9b00538

Fenimore C, Jones G (1961) Oxidation of ammonia in flames. J Phys Chem 65:298–303

Glarborg P, Hashemi H, Cheskis S, Jasper AW (2021) On the rate constant for NH_2+HO_2 and third-body collision efficiencies for NH_2+ H (+ M) and NH_2+ NH_2 (+M). J Phys Chem A 125(7):1505–1516. https://doi.org/10.1021/acs.jpca.0c11011

Gotama GJ, Hayakawa A, Okafor EC, Kanoshima R, Hayashi M, Kudo T, Kobayashi H (2022) Measurement of the laminar burning velocity and kinetics study of the importance of the hydrogen recovery mechanism of ammonia/hydrogen/air premixed flames. Combust Flame 236:111753

Gülder O, Snelling D, Sawchuk R (1996) Influence of hydrogen addition to fuel on temperature field and soot formation in diffusion flames. Symp (Int) Combust 2351–2358

Han X, Wang Z, Costa M, Sun Z, He Y, Cen K (2019a) Experimental and kinetic modelling study of laminar burning velocities of NH_3/air, NH_3/H_2/air, NH_3/CO/air and NH_3/CH_4/air premixed flames. Combust Flame 206:214–226. https://doi.org/10.1016/j.combustflame.2019.05.003

Han X, Wang Z, Costa M, Sun Z, He Y, Cen K (2019b) Experimental and kinetic modelling study of laminar burning velocities of NH3/air, NH3/H2/air, NH3/CO/air and NH3/CH4/air premixed flames. Combust Flame 206:214–226. https://doi.org/10.1016/j.combustflame.2019.05.00

Han X, Wang Z, He Y, Zhu Y, Cen K (2020) Experimental and kinetic modeling study of laminar burning velocities of NH3/syngas/air premixed flames. Combust Flame 213:1–13

Hayakawa A, Goto T, Mimoto R, Kudo T, Kobayashi H (2015a) NO formation/reduction mechanisms of ammonia/air premixed flames at various equivalence ratios and pressures. Mech Eng J 2:14–00402

Hayakawa A, Goto T, Mimoto R, Arakawa Y, Kudo T, Kobayashi H (2015b) Laminar burning velocity and Markstein length of ammonia/air premixed flames at various pressures. Fuel 159:98–106. https://doi.org/10.1016/j.fuel.2015.06.070

Hayakawa A, Goto T, Mimoto R, Kudo T, Kobayashi H (2015) NO formation/reduction mechanisms of ammonia/air premixed flames at various equivalence ratios and pressures. Mech Eng J 14–00402. https://doi.org/10.1299/mej.14-00402

He X, Shu B, Nascimento D, Moshammer K, Costa M, Fernandes R (2019) Auto-ignition kinetics of ammonia and ammonia/hydrogen mixtures at intermediate temperatures and high pressures. Combust Flame 206:189–200. https://doi.org/10.1016/j.combustflame.2019.04.050

Ichikawa A, Hayakawa A, Kitagawa Y, Somarathne KDKA, Kudo T, Kobayashi H (2015) Laminar burning velocity and Markstein length of ammonia/hydrogen/air premixed flames at elevated pressures. Int J Hydrogen Energy 40:9570–9578

Issayev G, Giri BR, Elbaz AM, Shrestha KP, Mauss F, Roberts WL, Farooq A (2022) Ignition delay time and laminar flame speed measurements of ammonia blended with dimethyl ether: a promising low carbon fuel blend. Renew Energy 181:1353–1370

Kim HK, Ku JW, Ahn YJ, Kim YH, Kwon OC (2021) Effects of O_2 enrichment on NH3/air flame propagation and emissions. Int J Hydrogen Energy 46(46):23916–23926

Klippenstein SJ, Harding L, Ruscic B, Sivaramakrishnan R, Srinivasan N, Su M-C et al (2009) Thermal decomposition of NH_2OH and subsequent reactions: ab initio transition state theory and reflected shock tube experiments. J Phys Chem A 113(38):10241–10259. https://doi.org/10.1021/jp905454k

Kobayashi H, Hayakawa A, Somarathne KDKA, Okafor EC (2019a) Science and technology of ammonia combustion. Proc Combust Inst 37:109–133

Kobayashi H, Hayakawa A, Somarathne KDKA, Okafor EC (2019b) Science and technology of ammonia combustion. Proc Combust Inst 37(1):109–133. https://doi.org/10.1016/j.proci.2018.09.029

Koch E (1945) Ammonia–a fuel for motor buses. J Inst Pet 31:213–233

Konnov AA (2009) Implementation of the NCN pathway of prompt-NO formation in the detailed reaction mechanism. Combust Flame 156:2093–2105

Law CK (2006) Combustion physics, Cambridge University Press, p 347

Lhuillier C, Brequigny P, Lamoureux N, Contino F, Mounaïm-Rousselle C (2020) Experimental investigation on laminar burning velocities of ammonia/hydrogen/air mixtures at elevated temperatures. Fuel 263:116653. https://doi.org/10.1016/j.fuel.2019.116653

Li J, Huang H, Kobayashi N, He Z, Nagai Y (2014) Study on using hydrogen and ammonia as fuels: combustion characteristics and NOx formation Int. J Energy Res 38:1214–1223

Li J, Huang H, Kobayashi N, He Z, Osaka Y, Zeng T (2015) Numerical study on effect of oxygen content in combustion air on ammonia combustion. Energy 93:2053–2068

Li R, Konnov AA, He G, Qin F, Zhang D (2019) Chemical mechanism development and reduction for combustion of NH_3/H_2/CH_4 mixtures. Fuel 257:116059

Lindstedt RP, Lockwood FC, Selim MA (1995) A detailed kinetic study of ammonia oxidation. Combust Sci Technol 108(4–6):231–254

Liu Q, Chen X, Huang J, Shen Y, Zhang Y, Liu Z (2019) The characteristics of flame propagation in ammonia/oxygen mixtures. J Hazard Mater 363:187–196. https://doi.org/10.1016/j.jhazmat.2018.09.073

Maclean DI, Wagner HG (1967) The structure of the reaction zones of ammonia-oxygen and hydrazine-decomposition flames. Proc Combust Inst 11:871–878

Mathieu O, Petersen EL (2015) Experimental and modeling study on the high temperature oxidation of ammonia and related NOx chemistry. Combust Flame 162(3):554–570. https://doi.org/10.1016/j.combustflame.2014.08.022

Medhat M, Yehia M, Khalil A, Franco MC, Rocha RC (2021) A numerical prediction of stabilized turbulent partially premixed flames using ammonia/hydrogen mixture. J Adv Res Fluid Mech Therm Sci 87(3):113–133

Mei B, Zhang X, Ma S, Cui M, Guo H, Cao Z, Li Y (2019) Experimental and kinetic modeling investigation on the laminar flame propagation of ammonia under oxygen enrichment and elevated pressure conditions. Combust Flame 210:236–246

Mendiara T, Glarborg P (2009) Ammonia chemistry in oxy-fuel combustion of methane. Combust Flame 156(10):1937–1949

Metghalchi M, James CK (1982) Burning velocities of mixtures of air with methanol, isooctane, and indolene at high pressure and temperature. Combust Flame 48:191–210

Michael J, Sutherland J, Klemm R (1986) Rate constant for the reaction, atomic hydrogen+ ammonia, over the temperature range, 750–1777 K. J Phys Chem 90(3):497–500

Miller JA, Bowman CT (1989) Mechanism and modeling of nitrogen chemistry in combustion. Prog Energy Combust Sci 15(4):287–338

Miller JA, Smooke MD, Green RM, Kee RJ (1983) Kinetic modeling of the oxidation of ammonia in flames. Combust Sci Technol 34(1–6):149–176

Montgomery MJ, Kwon H, Dreyer JA, Xuan Y, McEnally CS, Pfefferle LD (2021) Effect of ammonia addition on suppressing soot formation in methane co-flow diffusion flames. Proc Combust Inst 38(2):2497–2505

Nakamura H, Hasegawa S, Tezuka T (2017) Kinetic modeling of ammonia/air weak flames in a micro flow reactor with a controlled temperature profile. Combust Flame 185:16–27

Ni S, Zhao D, You Y, Huang Y, Wang B, Su Y (2021) NOx emission and energy conversion efficiency studies on ammonia-powered micro-combustor with ring shaped ribs in fuel-rich combustion. J Clean Prod 320:128901

Nozari H, Karabeyo˘glu A (2015) Numerical study of combustion characteristics of ammonia as a renewable fuel and establishment of reduced reaction mechanisms. Fuel 159:223–233

Okafor EC, Naito Y, Colson S, Ichikawa A, Kudo T, Hayakawa A, Kobayashi H (2018) Experimental and numerical study of the laminar burning velocity of CH4–NH3–air premixed flames. Combust Flame 187:185–198

Okafor EC, Naito Y, Colson S, Ichikawa A, Kudo T, Hayakawa A, Kobayashi H (2019) Measurement and modelling of the laminar burning velocity of methane-ammonia-air flames at high pressures using a reduced reaction mechanism. Combust Flame 204:162–175

Otomo J, Koshi M, Mitsumori T, Iwasaki H, Yamada K (2018) Chemical kinetic modeling of ammonia oxidation with improved reaction mechanism for ammonia/air and ammonia/hydrogen/air combustion. Int J Hydrogen Energy 43(5):3004–3014

Pearse RWB, Gaydon AG (1976) The identification of molecular spectra, Fourth Ed., John Wiley & Sons, Inc

Pfahl UJ, Ross MC, Shepherd JE, Pasamehmetoglu KO, Unal C (2000) Flammability limits, ignition energy, and flame speeds in H_2–CH_4–NH_3–N_2O–O_2–N_2 mixtures. Combust Flame 123:140–158

Prasad RK, Agarwal AK (2021) Experimental evaluation of laser ignited hydrogen enriched compressed natural gas fueled supercharged engine. Fuel 289(1):119788

Renard C, Dias V, Van Tiggelen P, Vandooren J (2009) Flame structure studies of rich ethylene–oxygen–argon mixtures doped with CO_2, or with NH_3, or with H_2O. Proc Combust Inst 32(1):631–637

Rocha RC, Ramos CF, Costa M, Bai X-S (2019) Combustion of NH_3/CH_4/Air andNH_3/H_2/air mixtures in a porous burner: experiments and kinetic modeling. Energy Fuels 33(12):12767–12780

Ronney PD (1988) Effect of chemistry and transport properties on near-limit flames at microgravity. Combust Sci Technol 59:123–141

Samu V, Varga T, Rahinov I, Cheskis S, Tur´anyi T (2018) Determination of rate parameters based on NH_2 concentration profiles measured in ammonia-doped methane–air flames. Fuel 212:679–83. https://doi.org/10.1016/j.fuel.2017.10.019

Shrestha KP, Seidel L, Zeuch T, Mauss F (2018) Detailed kinetic mechanism for the oxidation of ammonia including the formation and reduction of nitrogen oxides. Energy Fuels 32(10):10202–10217

Shrestha KP, Lhuillier C, Barbosa AA, Brequigny P, Contino F, Mounaïm-Rousselle C et al (2021) An experimental and modeling study of ammonia with enriched oxygen content and ammonia/hydrogen laminar flame speed at elevated pressure and temperature. Proc Combust Inst 38(2):2163–2174. https://doi.org/10.1016/j.proci.2020.06.197

Shrestha KP, Giri BR, Elbaz AM, Issayev G, Roberts WL, Seidel L, Mauss, Farooq A (2022) A detailed chemical insights into the kinetics of diethyl ether enhancing ammonia combustion and the importance of NOx recycling mechanism. Fuel Commun 100051

Shu B, He X, Ramos CF, Fernandes RX, Costa M (2021) Experimental and modeling study on the auto-ignition properties of ammonia/methane mixtures at elevated pressures. Proc Combust Inst 38(1):261–268

Singh AS, Mohapatra S, Boyapati R, Elbaz AM, Dash SK, Roberts WL, Reddy VM (2021) Chemical kinetic modeling of the autoignition properties of ammonia at low–intermediate temperature and high pressure using a newly proposed reaction mechanism. Energy Fuels 35(16):13506–13522

Skreiberg Ø, Kilpinen P, Glarborg P (2004) Ammonia chemistry below 1400 K under fuel-rich conditions in a flow reactor. Combust Flame 136(4):501–518

Smith GP, Golden DM, Frenklach M, Moriarty NW, Eiteneer B, Goldenberg M et al GRI Mech 3.0. Gas research institute. Accessed 24 May 2024. http://www.me.berkeley.edu/gri_mech/

Song Y, Hashemi H, Christensen JM, Zou C, Marshall P, Glarborg P (2016) Ammonia oxidation at high pressure and intermediate temperatures. Fuel 181:358–365

Stagni A, Cavallotti C, Arunthanayothin S, Song Y, Herbinet O, Battin- Leclerc F, Faravelli T (2020) An experimental, theoretical and kinetic-modeling study of the gas-phase oxidation of ammonia. React Chem and Eng 5(4):696–711

Sumathi R, Sengupta D, Nguyen MT (1998) Theoretical study of the H_2+ NO and related reactions of [H_2NO] isomers. J Phys Chem A 102(18):3175–3183. https://doi.org/10.1021/jp9804953

Takeishi H, Hayashi J, Kono S, Arita W, Iino K, Akamatsu F (2015) Characteristics of ammonia/N2/O2 laminar flame in oxygen-enriched air condition. Trans JSME 81(824):14–00423

Takizawa K, Takahashi A, Tokuhashi K, Kondo S, Sekiya A (2008) Burning velocity measurements of nitrogen-containing compounds. J Hazard Mater 155:144–152

Tian Z, Zhang L, Li Y, Yuan T, Qi F (2009) An experimental and kinetic modelling study of a premixed nitromethane flame at low pressure. Proc Combust Inst 32(1):311–318

UNCTAD (2019) Review of maritime transport

Verkamp FJ, Hardin MC, Williams JR (1967) Ammonia combustion properties and performance in gas-turbine burners. Symp (Int) Combust 985–992

Wang D, Ji C, Wang Z, Wang S, Zhang T, Yang J (2020) Measurement of oxy-ammonia laminar burning velocity at normal and elevated temperatures. Fuel 279:118425. https://doi.org/10.1016/j.fuel.2020.118425

Wang S, Elbaz AM, Wang Z, Roberts WL (2021a) The effect of oxygen content on the turbulent flame speed of ammonia/oxygen/nitrogen expanding flames under elevated pressures. Combust Flame 232:111521. https://doi.org/10.1016/j.combustflame.2021.111521

Wang Z, Han X, He Y, Zhu R, Zhu Y, Zhou Z et al (2021b) Experimental and kinetic study on the laminar burning velocities of NH_3 mixing with CH_3OH and C_2H_5OH in premixed flames. Combust Flame 229:111392. https://doi.org/10.1016/j.combustflame.2021.02.038

Xiao H, Howard M, Valera-Medina A, Dooley S, Bowen PJ (2016) Study on reduced chemical mechanisms of ammonia/methane combustion under gas turbine conditions. Energy Fuels 30(10):8701–8710

Yang Y, Zheng S, Sui R, Lu Q (2023) Impact of ammonia addition on soot and NO/N_2O formation in methane/air co-flow diffusion flames. Combust Flame 247:112483

Zakaznov FZ, Kursheva LA, Felina ZI (1978) Determination of normal flame velocity and critical diameter of flame extinction in ammonia-air mixture. Combust Explos. Shock Waves 14:710–713

Zhang Z, Li J, Tian J, Dong R, Zou Z, Gao S, Tan D (2022) Performance, combustion and emission characteristics investigations on a diesel engine fueled with diesel/ethanol/n-butanol blends. Energy 249:123733

Zhou M, Yan F, Ma L, Jiang P, Wang Y, Chung SH (2022) Chemical speciation and soot measurements in laminar counterflow diffusion flames of ethylene and ammonia mixtures. Fuel 308:122003

Zincir B, Shukla PC, Agarwal AK (2023) Introduction to decarbonization of maritime transport. In: Zincir B, Shukla PC, Agarwal AK (eds) Decarbonization of maritime transport. Energy, environment, and sustainability. Springer, Singapore. https://doi.org/10.1007/978-981-99-1677-1_1

Chapter 6
Decoding the Challenge of Ammonia Combustion: A Hybrid Approach Considering CFD and AI

Aswitha Tadepalli, Richie Shaji Mathew, Raja Banerjee, and Kishalay Mitra

Abstract The transition to carbon–neutral fuels is crucial considering the alarming threats of climate change and global warming. Though hydrogen (H_2) is the best carbon–neutral fuel owing to its highest calorific value, its storage, transportation, and the possibility of embrittlement pose many challenges for its usage. Hence, due to its high gravimetric hydrogen, ammonia (NH_3) is currently being exploited as the H_2 carrier. However, the usage of NH_3 directly as a fuel is not effective owing to its low laminar flame speed, low flame characteristics and high nitrogen oxide (NO_x) emissions. Thus, it is generally blended with H_2 and carbon fuels to improve its flame characteristics. The main aim of the study is thus to find the optimum blend of NH_3, natural gas (CH_4), n-dodecane and H2 that minimizes NH3 slip along with minimization of CO and NO_x emissions. The combustion mechanism of this blend fuel involves 242 species and 1769 reactions including both NOx and soot sub-mechanisms. Considering such a mechanism in a Computational Fluid Dynamics (CFD) framework, first a high-fidelity model is constructed to understand the several species behaviour under the combustion process. However, finding the optimum parameters using the high-fidelity combustor model is computationally very expensive due to the iterative nature of optimization modules. Therefore, this study focuses on developing Artificial Neural Network (ANN) models that emulate the CFD model and finally solving the afore-mentioned optimization formulation using the ANN surrogates. The data points required for developing the ANN model were generated through SOBOL sampling using four parameters, viz., wattage, NH_3 composition in fuel, pressure and air preheat temperature, where CFD simulations

A. Tadepalli · K. Mitra
Department of Chemical Engineering, Indian Institute of Technology Hyderabad, Kandi, Sangareddy, Telangana, India

R. S. Mathew · R. Banerjee (✉)
Department of Mechanical and Aerospace Engineering, Indian Institute of Technology Hyderabad, Kandi, Sangareddy, Telangana, India
e-mail: rajabanerjee@mae.iith.ac.in

K. Mitra
Department of Climate Change, Department of Artificial Intelligence, Indian Institute of Technology Hyderabad, Kandi, Sangareddy, Telangana, India

© The Author(s), under exclusive license to Springer Nature Singapore Pte Ltd. 2026
H. Valera and A. K. Agarwal (eds.), *Ammonia and Low Carbon Combustion Engine Technologies for De-Fossilizing Transport Sector*, Energy, Environment, and Sustainability, https://doi.org/10.1007/978-981-96-6628-7_6

were performed. Using the wattage, NH_3 composition, CH_4 composition in fuel, pressure and air preheat temperature as inputs, and compositions of CO, NH_3 and NO_x in the flue gas as outputs, three ANN surrogates were developed. Using these surrogates, the optimization problem was solved using the Non-dominated Sorting Genetic Algorithm (NSGA II). Based on the requirement, the fuel blend with high NH_3 composition and low CH_4 composition or the one with low NH_3 composition and high CH_4 composition can thus be selected from the obtained Pareto solutions.

Keywords Ammonia combustion · CFD modelling · AI/ML · Multi-objective optimization

6.1 Introduction

One of the United Nation's Sustainable Development Goals (SDGs) includes the goal on affordable and clean energy, known as SDG 7 (https://sdgs.un.org/goals), that aims to ensure access to affordable, reliable, sustainable and modern energy for all. Modern society is very energy dependent and combustion of traditional fossils like coal, oil and natural gas for energy and transportation releases large amounts of CO_2 into the atmosphere. CO_2 and other greenhouse gases like CH_4 is contributing to a rapid rise in the average global temperature that is expected to contribute towards climate change, rising sea levels, disruption of the ecosystem and bio-diversity and human health (Al-Ghussain 2019). These challenges necessitate research in carbon neutral fuels such as biofuels, hydrogen and synthetic fuels, aim to produce net-zero carbon dioxide (CO_2) emissions. Combustion research in such fuels also helps in decrease of air pollution, leading to better air quality and associated health benefits.

CO_2 emissions during combustion are directly proportional to the H/C ratio in the fuel blend (Heywood 2018). As the ratio of hydrogen increases in the fuel, CO_2 emission will decrease. The best-case scenario is when the fuel consists of only hydrogen and there will no CO_2 emission. Only water vapour will be emitted during the combustion process. However, hydrogen storage and transportation are a challenge due to its low density. Therefore, its energy content per unit volume is low (Dimitriou and Tsujimura 2017). It has to be stored in tanks at very high pressures which poses new challenges. Ammonia is another candidate fuel where there is no carbon. It can be stored in liquid form at 10 bar pressure (Dimitriou and Javaid 2020). However, its combustion characteristics like low laminar flame speed, long ignition delay and propensity to form NO_x makes it a difficult fuel to burn. Therefore, researchers are exploring the option of blending it with other fossil fuels like natural gas and study its combustion characteristics. Various parameters affect its combustion characteristics like blending ratio, equivalence ratio, system pressure and premixing ratio with air. As many of these have competing effects of combustor performance in terms of fuel conversion efficiency and emissions, optimization of combustion operating parameters is required. Lately. Artificial Intelligence and Machine Learning methods have been used to perform optimization of various industrial processes (Mosavi et al. 2019; Lee et al. 2018; Bikmukhametov and Jäschke 2020; Shang and You 2019).

Machine Learning techniques are now increasingly being used to analyse various problems in fluid mechanics (Brunton et al. 2020; Duraisamy 2021). It has been used to study turbulent flows (Narasimhan et al. 2021), external aerodynamics (Provost and Eldredge 2021), high speed flows (Mao et al. 2020) and detonation (Mendible et al. 2021) among other types of flows. Various aspects of combustion have been studied using Machine Learning methods. That includes combustion chemical kinetics development (Ryu 2021) and its reduction (Wan et al. 2020), accelerated chemistry calculations in combustion studies (Mehl and Aubagnac-Karkar 2023) and optimizing combustor operation (Debiagi et al. 2020; Probst et al. 2019; Jabbr et al. 2020).

Recently machine learning (ML) has gained huge momentum to solve several complex problems in diverse areas such as industry, agriculture, medical diagnosis. Ravi Kiran et al. used Recurrent Neural Networks to model the industrial grinding operation (Inapakurthi et al. 2020). Keerthi et al. exploited Long Short-Term Memory networks for better wind forecasting (Pujari et al. 2023). To detect pneumonia using X-rays Nahid et al. proposed a model based on Convolutional Neural Network (Nahid et al. 2020).

The main aim of this study is to perform optimization for the combustion process of dodecane/CH_4/H_2/NH_3 fuel blend. Due to toxic nature, minimization of CO, unburnt NH_3 and NO_x emissions in the flue gas is badly needed and hence targeted. To attain this objective, firstly, a CFD model is simulated to understand the combustion process of the dodecane/CH_4/H_2/NH_3 fuel blend more accurately. Owing to the iterative nature of the optimizers, optimization of the combustion process using CFD model would be tremendously time-consuming given each of the function evaluation using CFD model is time expensive. Thus, in this study, an Artificial Neural Network (ANN) based surrogate is developed to emulate the CFD model behaviour. Using this ANN surrogate, the multi-objective optimization problem of minimizing CO, NH_3 and NO_x emissions in the flue gas is solved using the NSGA II algorithm.

The rest of the chapter is organized as follows. In Sect. 6.2, the governing equations used for the development of the CFD model, the working principles of ANN and NSGA II are briefly described. In Sect. 6.3, the CFD model and its validation results are presented. In Sect. 6.4, the ANN results and the optimization results are discussed and finally in Sect. 5 the conclusions of the study are collated.

6.2 Methodology

6.2.1 Design of Experiments

To generate the dataset required for the development of the ML model, the SOBOL sequence is employed to ensure uniform distribution of the Design of Experiments (DoE) within the entire input parameter space (Burhenne et al. 2011). The parameters

Table 6.1 Lower and upper bounds of the parameters used to generate the DoE

Parameter	Lower bound	Upper bound
Wattage (KW)	5	20
Pressure (bar)	1	50
NH_3 in the fuel blend (%)	10	85
Air pre heat temperature (K)	298	600

that are varied during the generation of DoE include wattage, pressure, air preheat temperature and percentage of NH_3 in the dodecane/CH_4/H_2/NH_3 fuel blend. In this study, the dodecane % in the fuel blend was fixed at 10 that of H2 as 5% of NH_3 and the CH_4 constitutes the remaining percentage. Thus, only the NH_3% is varied for the gas mixtures and the bounds of the parameters within which the 200 DoE is generated are shown in Table 6.1. To validate the uniformity of the generated points, scatter plots of the generated DoE are generated and are shown in Fig. 6.1.

The wattage and pressure ranges are limited by experimental feasibility. Pure ammonia (NH_3) has low combustion characteristics, so it is blended with methane (CH_4) to improve performance. However, increasing CH_4 raises carbon emissions. Pre-heating the air reduces ignition delay of the fuel mixture but decreases combustor efficiency due to the negative work involved.

6.2.2 Combustion Model

Flamelet model (Peters 1999) was used to simulate turbulent combustion in this work. It is suitable for non-premixed type combustor configuration where the combustion kinetics are fast and the time scales for turbulence and chemical kinetics can be separated. Favre averaged conservation of mass, momentum and energy equations are solved in the flow domain. Turbulence modelling was performed using the Realizable $k - \varepsilon$ model. Additionally, two scalar transport equations are also solved while simulating combustion using flamelet method. One is the mean fraction equation and the second one is variance of the mixture fraction. Mixture fraction is defined as

$$f = \frac{Z_i - Z_{i,ox}}{Z_{i,\text{fuel}} - Z_{i,ox}}$$

where Z_i is the elemental mass fraction of element, i. It takes the value of 1 in the fuel stream and 0 in the oxidizer stream, and it monotonically changes across the flame front. The variation in the mixture fraction due to turbulence is represented by a probability density function and it has been taken to be a β-function in this case.

A turbulent flame is assumed to be an ensemble of multiple laminar flamelets, while using the flamelet model. It is also assumed that the gradients normal to the flame front is significantly higher than in the tangential direction and therefore, a coordinate transformation can be done from the physical space to the mixture fraction

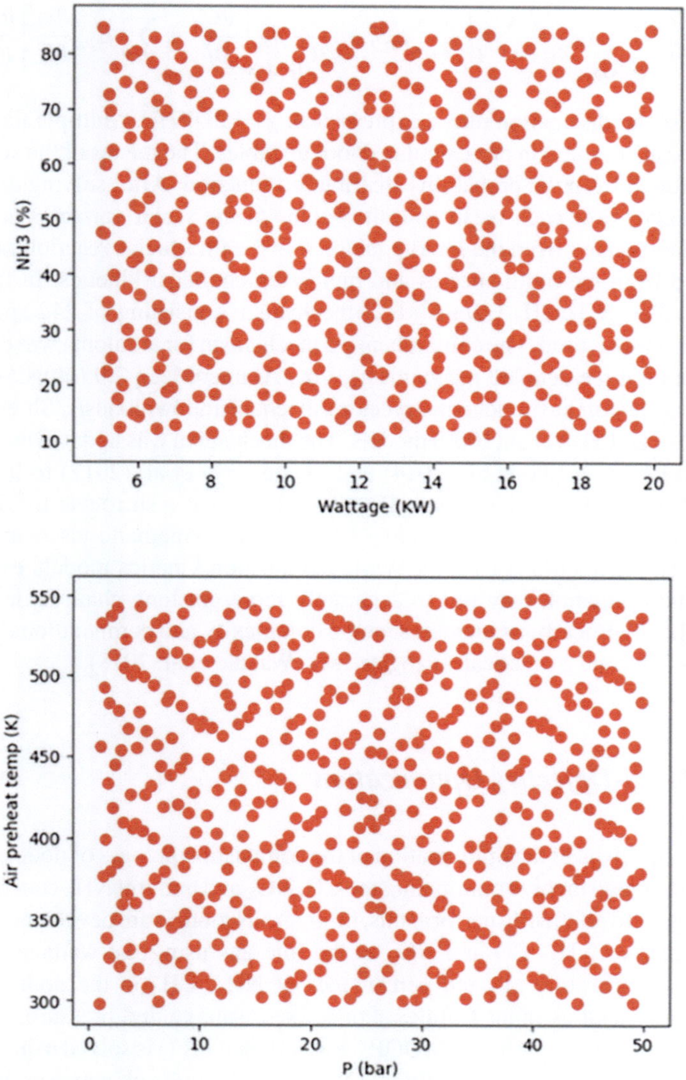

Fig. 6.1 Scatter plots of the points generated through the DoE for $NH_3\%$ and wattage, air preheat temperature and pressure

space. This results in 1-D equations of species and temperature of the form (Peters 1999),

$$\rho \frac{\partial y_i}{\partial t} = \frac{1}{2} \rho \chi \frac{\partial^2 y_i}{\partial f^2} + S_i$$

$$\rho\frac{\partial T}{\partial t} = \frac{1}{2}\rho\chi\frac{\partial^2 T}{\partial f^2} - \frac{1}{C_p}\sum_i h_i S_i + \frac{1}{2C_p}\rho\chi\left[\frac{\partial C_p}{\partial f} + \sum_i C_{p,i}\frac{\partial y_i}{\partial Z}\right]\frac{\partial T}{\partial f}$$

where the scalar dissipation rate is expressed as $\chi = 2D|\nabla \mathrm{f}|^2$. Multiple flamelets of varying χ can be stored in pre-calculated lookup tables. The means of the scalar variables are found from the probability distribution function. After solving the conservation equations for each mesh cell, the values of the scalar variables at that cell are then interpolated from the lookup table. S_i is the ith species reaction rate which is obtained from the multi-species and multi-step chemical kinetics. In this work, a semi-detailed NH_3/CH_4/H2/n-C_{12}H2$_6$ mechanism consisting of 242 species and 1769 reactions was used. The comprehensive mechanism for ammonia was developed by combining three detailed mechanisms, viz., AramcoMech 2.0 (https://c3.universityofgalway.ie/combustionchemistrycentre/mechanismdownloads/), Shrestha et al. (2021, 2018) and Tian et al.(2009) models. The mechanism was updated using mechanism developed by Luo et al. (2014) and Slavinskaya et al. (2012) to include the n-C12H26 and PAH mechanisms. n-Dodecane is used as a surrogate for diesel due to its similar fluid properties, including boiling point, kinematic viscosity, heating value, density and surface tension. While combustion kinetics models exist for n-heptane and n-butanol, they fail to accurately represent long-chain hydrocarbons. Using higher hydrocarbons increases model complexity and computational expense, making n-dodecane a practical surrogate fuel (Wakale et al. 2018).

6.2.3 Multi-Objective Optimization

The main objectives of the optimization of the combustion process of dodecane/CH_4/H2/NH_3 fuel blend involve minimizing the CO, NO_x and unburnt NH_3 concentrations in the flue gas. To efficiently perform this, firstly ANN models are developed to predict the flow rates of NH_3, CO and NO_x in the flue gas using the wattage, pressure, air preheat temperature (T) and percentages of NH_3, CH_4 in the dodecane/CH_4/H2/NH_3 fuel blend as input features. For a fixed wattage and pressure, the multi-objective optimization problem (MOOP) as mentioned in (1) is solved using the Non-dominated Sorting Genetic Algorithm II (NSGA-II). In multi-objective optimization, more than one objective is considered to be optimized simultaneously, which needs a trade-off among the objectives to be present. Due to this, in MOOP, there exists more than one solution compared to single solution in case of single objective optimization problem. The aim in solving a MOOP is, therefore, to find this set of trade-off solutions, known as the Pareto set of solutions. The NH_3% in the fuel blend and air preheat temperature are considered as the decision variables and using these variables, the fixed wattage, pressure and the CH_4% in the fuel blend (calculated as 90–1.05 × NH_3%), the flow rates of NH_3, CO and NO_x in the flue gas are predicted using the developed ANN models. Based on these predictions, the optimizer updates the values of the decision variables in every iteration such that the predictions, i.e.

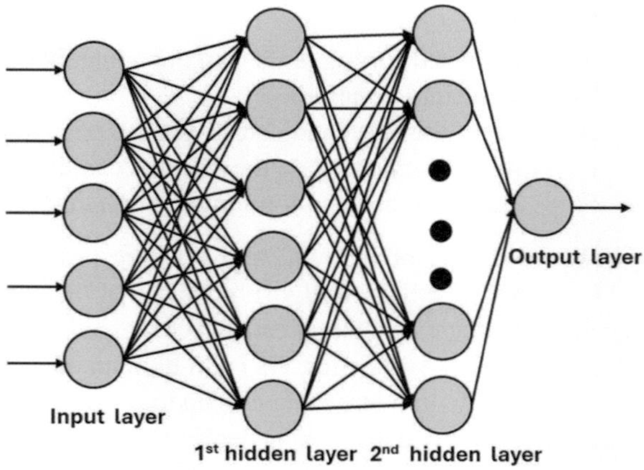

Fig. 6.2 Structure of a simple ANN with two hidden layers

the objective functions are minimized.

$$\min_{NH_3, T} NH_3, NO_x, CO \tag{6.1}$$

subject to

Flamelet model equations (as embedded in the ANN model).

6.2.3.1 Artificial Neural Networks

ANNs, inspired by the human brain structure and functionality, are designed to process and learn hidden patterns from huge amounts of data. ANNs can be used in complex tasks such as predictive modelling, image processing and natural language processing. They consist of several processing units called neurons that are interconnected across several layers (Bishop and Nasrabadi 2006). The structure of a simple ANN with five inputs and one output is shown in Fig. 6.2. The activation functions such as ReLU, sigmoid, tanh applied in the neurons are responsible for capturing the non-linearity present in the data. The neurons at one hidden layer are connected to those in the next hidden layer with certain weights and biases which are updated during the training process such that the error between true outputs and predicted outputs by the ANN is minimized (Miriyala et al. 2022).

6.2.3.2 Non-Dominated Sorting Genetic Algorithm II

NSGA II is an evolutionary optimization algorithm widely used for solving multi-objective optimization problems. The steps involved in the NSGA II are.

1. Random initialization of parent population of a specific size

2. Evaluation of fitness values
3. Non-dominated sorting of the parent population (along with the child population from the second iteration)
4. Selection of the best candidates from the population
5. Crossover and mutation on the selected candidates
6. Steps 2 to 5 are repeated for a maximum number of iterations (generations).

The flow sheet of the NSGA II working procedure is shown in Fig. 6.3 (Deb et al. 2002). The NSGA II algorithm is one of the popular techniques because of its capability to generate well-spread and converged Pareto solutions. This is because of its three main characteristics, viz., non-dominated sorting of the population to select the candidates with better rank, elitism to preserve the elite candidates of the parent population directly in the next generation and the crowding-distance-based selection to identify the best candidates among those with the same ranks (Verma et al. 2021).

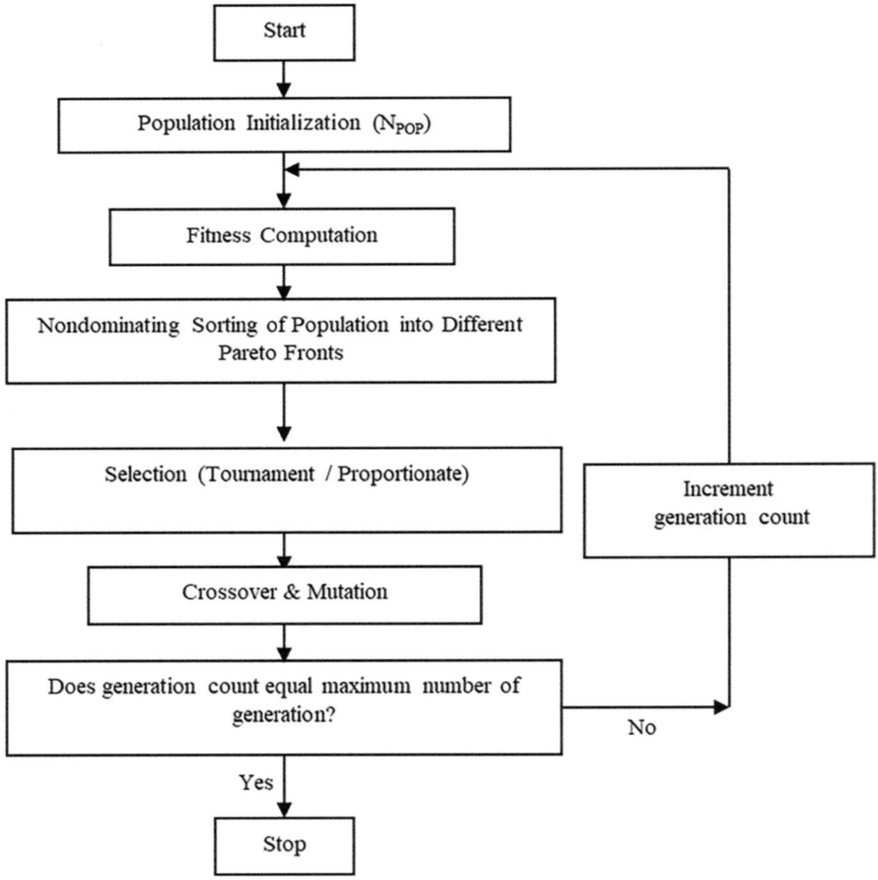

Fig. 6.3 Flowsheet of NSGA II

6.2.4 CFD Modelling and Validation

The geometric domain was adopted from the experimental setup used by Brookes and Moss (1999). The computational domain used in this work is the same as the one used by Garnayak et al. (2022).

The schematic of the experimental setup and the computational domain is shown in Fig. 6.4. The fuel consists of pure Methane (CH4), and the oxidizer is dry air. The fuel is issued through a nozzle of diameter (d) 4.7 mm with a mass flow rate of 10.3 g/min. Surrounding the fuel nozzle, co-flow air is supplied at 708 g/min. The fuel and co-flow air are supplied at 290 K, and the ambient pressure is 1 atm. The domain is a cylindrical tube with a diameter of 155 mm. The walls of the tube are considered to be adiabatic, and the outlet is open to the atmosphere. From the work performed by Garnayak et al. (2022), it was found that the Froude number was 0.592, indicating that the flow is a buoyancy-dominant flow. Hence, gravity is considered during the analysis. 2D axisymmetric analysis employing the $k-\varepsilon$ turbulence model was conducted with the corrected model constants used by Garnayak et al. (2022).

The validation of the CFD setup with the experiments and the numerical analysis for axial and radial distribution at r = 0.15 m, 0.25 m and 0.425 m is shown in Fig. 6.5. The numerical results from the current work were compared with the data reported by Brookes and Moss (1999) and the numerical results from the work of

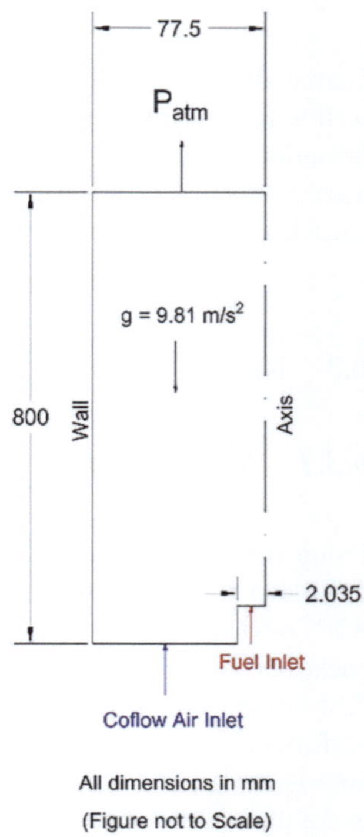

Fig. 6.4 Schematic view of the experimental and computational domain

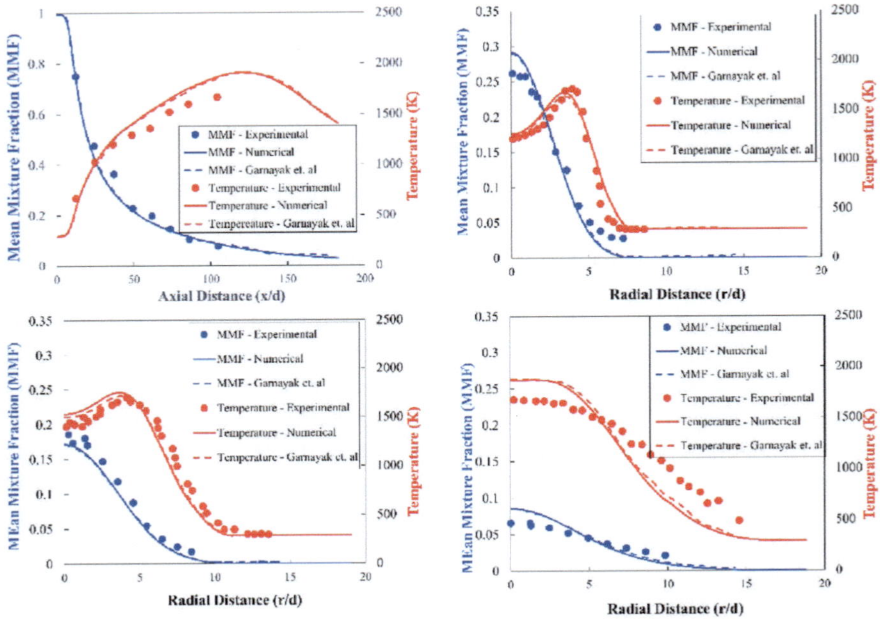

Fig. 6.5 Validation of the CFD; **a** Axial Distribution **b**, **c**, **d** Radial Distribution at r = 0.15 m, 0.25 m and 0.425 m respectively

Garnayak et al. (2022). Results from both the numerical simulations match very well with each and are almost superimposed on each other. Whereas there is some deviation in the temperature profiles reported in the experiments. This is because the numerical studies neglected radiative heat transfer effects due to the presence of soot particles and this assumption can be attributed to the observed deviation.

6.3 Results and Discussion

6.3.1 ANN Results

Using the dataset generated through the DoE, three ANN models are developed to predict the flow rates of NH3, CO and NOx in the flue gas respectively. The ANN modelling was performed in Python using pandas, scikit-learn and TensorFlow packages. 80% of the data, i.e. 160 points were used for training and validating the ANN models, and the rest 20%, i.e. 40 points were used for testing the model's performance on unseen data. The dataset was normalized to ensure that all the features contribute equally to the prediction. The ANN models were trained for 100 epochs using the training dataset and the best architectures were identified using the trial-and-error method. The best architectures for all three models and their corresponding

Table 6.2 The best architectures of the three ANN models for the prediction of CO, NH_3, NO_x flow rates and their corresponding R2 scores for training and testing

Output	Architecture	R2 training	R2 testing
CO	[5-6-16-1]	0.95	0.91
NH_3	[5-5-16-1]	0.93	0.92
NO_x	[5-6-16-1]	0.88	0.88

R2 scores for training and testing are presented in Table 6.2. Architecture [5–6-16–1] for CO as output model means that there are 2 intermediate layers with 6 and 16 nodes present, respectively, between the 5 inputs and 1 output variables.

From the table, it could be seen that the R2 scores for both training and testing data are high for all the models indicating both the effectiveness and the robustness of the models. To visualize the performance of the models, the parity plots are plotted and are shown in Fig. 6.6. From the plots, it can be seen that the model predictions agree very closely with the true values as most of the points are aligned across the 450-line indicating the highly accurate predictions of the models. It can be highlighted that only with a small dataset of 200 points, the developed ANN models have a proficient performance.

6.3.2 *Multi-Objective Optimization*

To visualize and validate the trade-off among the three objective functions, the contour plots of the flow rates of NH3, NOx and CO were created. To generate the contour plot, a mesh grid of 10,000 points was generated within the entire input space, and the flow rates were obtained at these points using the ANN models' predictions. The contours plotted for unburnt NH3, NOx and CO in the flue gas by considering wattage and NH3% in the fuel blend are shown in Fig. 7a–c, respectively.

From the above contours, it can be inferred that for a fixed wattage, as NH3% in the fuel blend increases, the NOx flow rate increases, while the CO and unburnt NH3 flow rates decrease. This indicates that although the CO and NH3 flow rates are favourable, the NOx flow rate is competing with these two. Thus, there is a clear trade-off among the CO, NH3 and NOx flow rates thereby indicating that the formulated MOOP is feasible. The MOOP was then solved at a fixed wattage of 10 KW, and pressure of 1 bar using the NSGA II framework in Python using the *pymoo* package. To ensure the reproducibility of the results, a random seed of 42 was set in the algorithm. The algorithm was run for 100 population and 100 generations. The Pareto front obtained after the 100th generation is shown in Fig. 6.8.

From the Pareto front, it can be concluded that the optimization runs were highly effective as the trade-off among the three competing objective functions can be visualized in the obtained Pareto front. Further, it can be seen that the obtained solutions are non-dominated, well-diversified and are also uniformly distributed across the

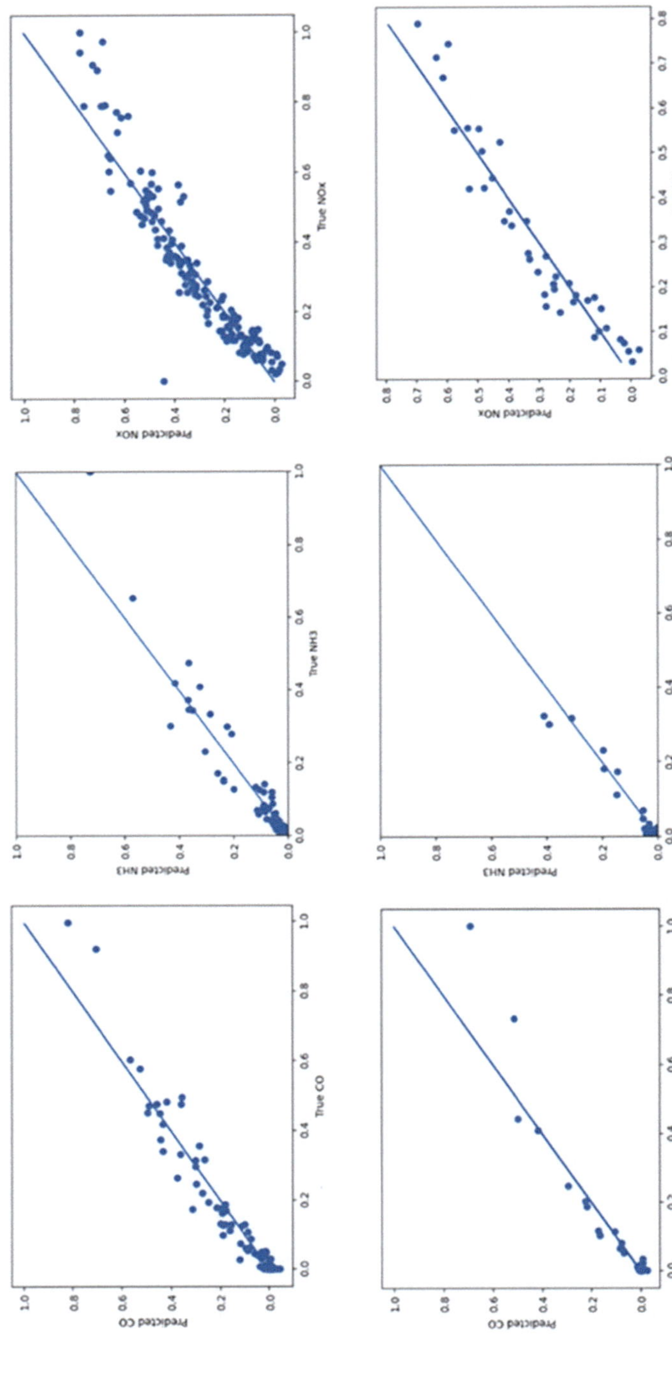

Fig. 6.6 Parity plots for CO, NH3 and NOx flowrates for training (top row) and testing data (bottom row)

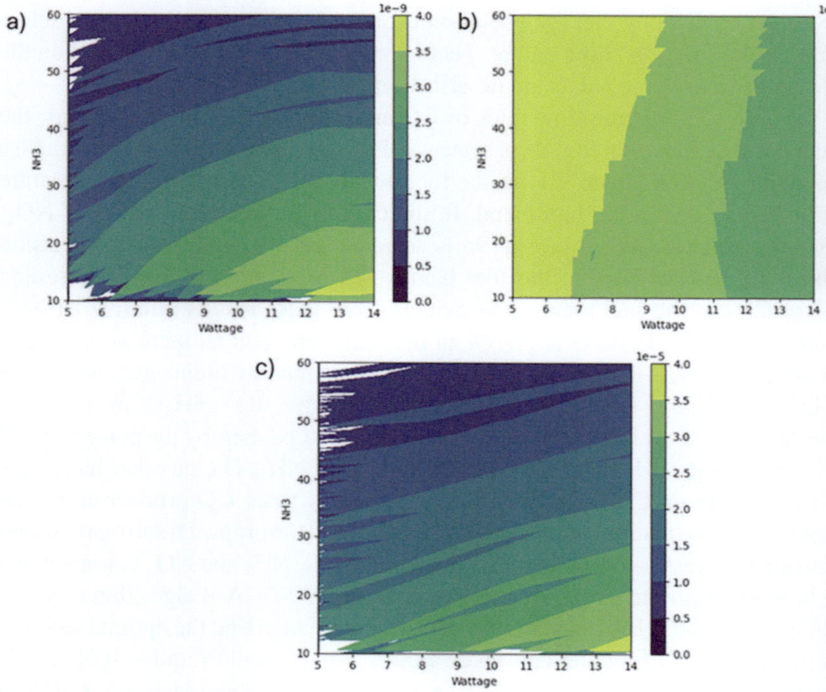

Fig. 6.7 Contour plots for NH_3, NO_x and CO flowrates plotted against NH_3% in the fuel blend and wattage

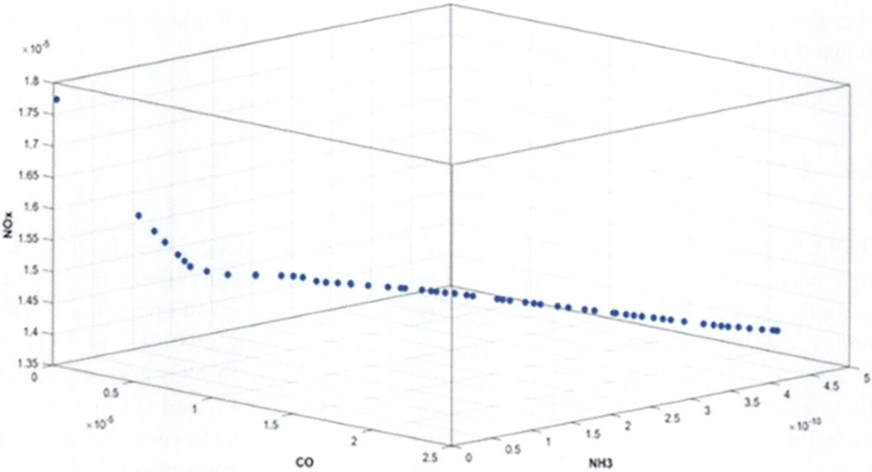

Fig. 6.8 Pareto front obtained from the NSGA II

entire domain indicating that each solution is optimal thereby facilitating the selection of the best solution based on the specific requirement. The results based on the selection of two extreme points are described in the following paragraph.

In the case of minimum flow rates of CO and unburnt NH_3 in the flue gas, the extreme point as shown in Fig. 9a is selected. The corresponding decision variables of this point are 45.67% of NH_3 in the fuel blend, 545 K air preheat temperature and 42.05% of CH_4 in the fuel blend. In the case of minimum flowrates of NO_x, the extreme point as shown in Fig. 9b is selected and its corresponding decision variables are 10% of NH_3 in the fuel blend, 545 K air preheat temperature and 79.5% of CH_4 in the fuel blend. It is evident from these results minimizing NO_x emissions favours a decrease in NH_3% in the fuel blend and an increase in CH_4% while minimizing CO favours a decrease in CH_4% in the fuel blend and an increase in NH3%. Analysing from the combustion perspective, less NH_3% in the blend implies less NH_3% available for reaction to produce NOx, thereby the percentage of NOx in the flue gas is less compared to the unreacted NH_3. On the other hand, less NH_3% implies more CH_4% in the fuel blend thereby more CO production by the reactions. Thus, the results obtained can be justified and the proposed surrogate-based optimization process can greatly help minimize the CO, NH_3 and NO_x compositions in the flue gas as per the target. It could be seen that the NSGA II algorithm required 10,000 function calls (100 population $\times$ 100 generations) to find the optimal solution set. Since each CFD function call takes around 1 h, it would require 10,000 h to complete the optimization process. Hence, surrogate based approach was followed in which only 200 CFD function calls were taken to generate the data and develop the surrogate (ANN) model. Thus, it requires only 200 h (for generating data) and around 5 h (for developing the ANN model and executing the NSGA II algorithm). Therefore, the time required for the optimization process has tremendously reduced to just 205 h.

6.4 Conclusions

Owing to the harmful threat of global warming and climate change, there is an immediate need to replace conventional fossil fuels with carbon–neutral fuels. Hydrogen, being the best choice for the replacement, poses many storage and operational challenges, causing difficulties in its utilization and necessitating the usage of NH_3 as an H2 carrier. However, due to the low flame speed and high NO_x emissions, the direct usage of NH_3 as the fuel is inefficient. Thus, lately, NH_3 and H2 fuel blends are being exploited to improve the combustion performance. However, because of their toxicity, the NO_x, CO, unburnt NH_3 emissions must be minimized. Therefore, in the study, the following conclusions are drawn.

- Firstly, the combustion performance of n-dodecane/CH_4/NH_3/H_2 fuel blend was studied using a CFD model. Since optimization using this time-expensive CFD model is tremendously time-consuming, ANN-based optimization was performed

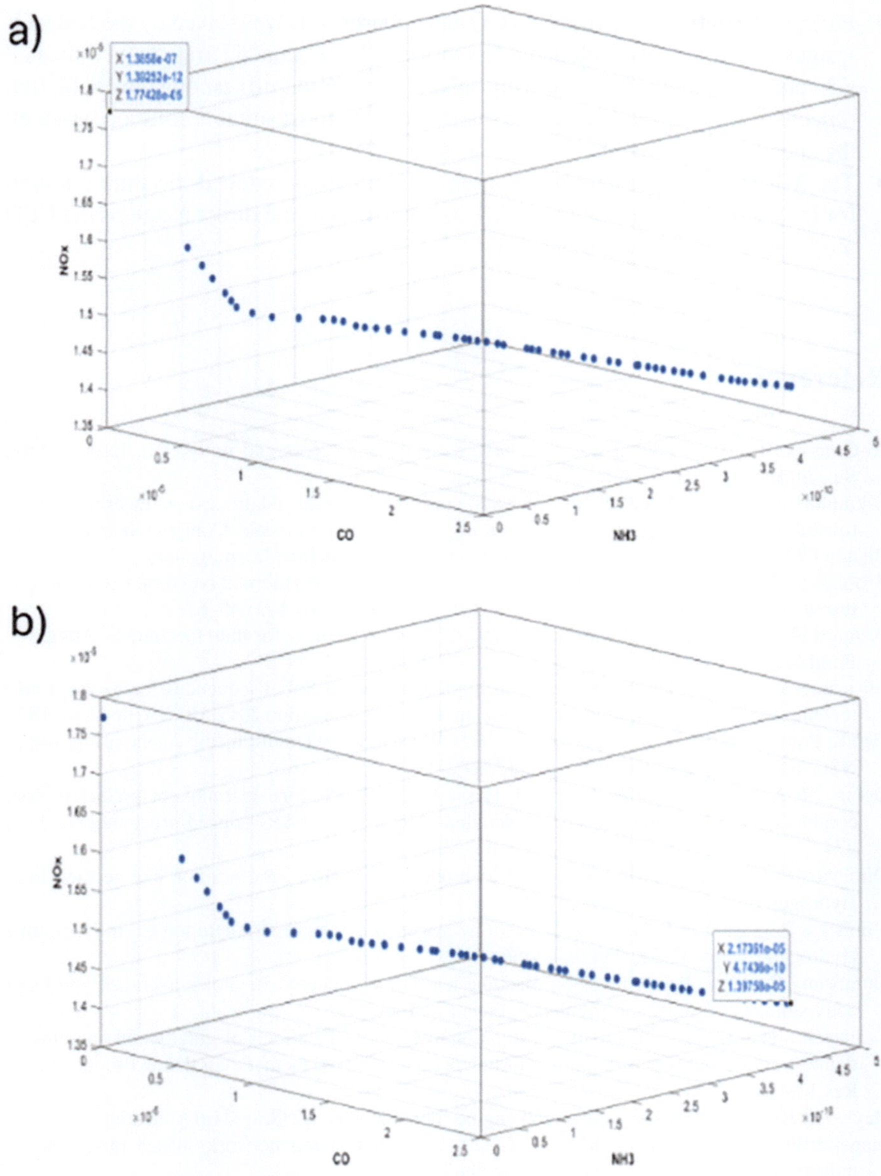

Fig. 6.9 Selection of two extreme points from the obtained Pareto front

in this study. Thus, ANN surrogates were developed to predict the flow rates of NO_x, CO and NH_3 in the flue gas using 200 data points generated from the CFD model. The high R2 scores for both the training and testing data validated the excellent efficiency and robustness of the ANN model.

- Further, the optimization of NO_x, CO and unburnt NH_3 was solved by the NSGA II framework using the developed ANN models. The obtained Pareto front indicated a diverse set of solutions effectively balancing the trade-offs among the conflicting objectives thereby facilitating the selection of the most suitable solution based on the specific requirement.
- The ML surrogate based optimization technique has reduced the time required for the optimization process by 98% as compared to the direct usage of the CFD model in the optimization process.

References

Al-Ghussain L (2019) Global warming: review on driving forces and mitigation. Environ Prog Sustain Energy 38(1):13–21

Bikmukhametov T, Jäschke J (2020) Combining machine learning and process engineering physics towards enhanced accuracy and explainability of data-driven models. Comput Chem Eng 138

Bishop CM, Nasrabadi NM (2006) Pattern recognition and machine learning, 4(4)

Brookes SJ, Moss JB (1999) Measurements of soot production and thermal radiation from confined turbulent jet diffusion flames of methane. Combust Flame 116(1–2):49–61

Brunton SL, Noack BR, Koumoutsakos P (2020) Machine learning for fluid mechanics. Annu Rev Fluid Mech 52:477–508

Burhenne S, Jacob D, Henze GP (2011) Sampling based on Sobol'sequences for Monte Carlo techniques applied to building simulations. In: Building simulation 2011, IBPSA, pp 1816–1823

Deb K, Pratap A, Agarwal S, Meyarivan T (2002) A fast and elitist multiobjective genetic algorithm: NSGA-II. IEEE Trans Evol Comput 6(2):182–197

Debiagi P, Nicolai H, Han W, Janicka J, Hasse C (2020) Machine learning for predictive coal combustion CFD simulations—from detailed kinetics to HDMR reduced-order models. Fuel 274

Dimitriou P, Javaid R (2020) A review of ammonia as a compression ignition engine fuel. Int J Hydrogen Energy 45(11):7098–7118

Dimitriou P, Tsujimura T (2017) A review of hydrogen as a compression ignition engine fuel. Int J Hydrogen Energy 42(38):24470–24486

Duraisamy K (2021) Perspectives on machine learning-augmented Reynolds-averaged and large eddy simulation models of turbulence. Phys Rev Fluids 6(5)

Garnayak S et al (2022) Numerical investigation to evaluate the effects of gravity and pressure on flame structure and soot formation of turbulent non-premixed methane-air flame. Propuls Power Res 11(4):473–495

Heywood JB (2018) Internal combustion engine fundamentals. McGraw-Hill Education

Inapakurthi RK, Miriyala SS, Mitra K (2020) Recurrent neural networks based modelling of industrial grinding operation. Chem Eng Sci 219

Jabbr AI, Gaja H, Koylu UO (2020) Multi-objective optimization of operating parameters for a H2/Diesel dual-fuel compression-ignition engine. Int J Hydrogen Energy 45(38):19965–19975

Lee JH, Shin J, Realff MJ (2018) Machine learning: overview of the recent progresses and implications for the process systems engineering field. Comput Chem Eng 114:111–121

Luo Z et al (2014) Development and validation of an n-dodecane skeletal mechanism for spray combustion applications. Combust Theor Model 18(2):187–203

Mao Z, Jagtap AD, Karniadakis GE (2020) Physics-informed neural networks for high-speed flows. Comput Methods Appl Mech Eng 360

Mehl C, Aubagnac-Karkar D (2023) On-the-fly accuracy evaluation of artificial neural networks and hybrid method to improve the robustness of neural network accelerated chemistry solving. Phy Fluids 35(6)

Mendible A, Koch J, Lange H, Brunton, Kutz JN (2021) Data-driven modeling of rotating detonation waves. Phys Rev Fluids 6(5)

Miriyala SS, Pujari KN, Naik S, Mitra K (2022) Evolutionary neural architecture search for surrogate models to enable optimization of industrial continuous crystallization process. Powder Technol 405

Mosavi A, Shamshirband S, Salwana E, Chau K-W, Tah JHM (2019) Prediction of multi-inputs bubble column reactor using a novel hybrid model of computational fluid dynamics and machine learning. Eng Appl Comput Fluid Mech 13(1):482–492

Nahid A-A et al (2020) A novel method to identify pneumonia through analyzing chest radiographs employing a multichannel convolutional neural network. Sensors (Switzerland) 20(12):1–18

Narasimhan G, Meneveau C, Zaki TA (2021) Large eddy simulation of transitional channel flow using a machine learning classifier to distinguish laminar and turbulent regions. Phys Rev Fluids 6(7)

NUIG Combustion Chemistry Center. https://c3.universityofgalway.ie/combustionchemistrycentre/mechanismdownloads/

Peters N (1999) The turbulent burning velocity for large-scale and small-scale turbulence. J Fluid Mech 384:107–132

Probst DM et al (2019) Evaluating optimization strategies for engine simulations using machine learning emulators. J Eng Gas Turbine Power 141(9)

Le Provost M, Eldredge JD (2021) Ensemble Kalman filter for vortex models of disturbed aerodynamic flows. Phys Rev Fluids 6(5)

Pujari KN, Miriyala SS, Mittal P, Mitra K (2023) Better wind forecasting using evolutionary neural architecture search driven green deep learning. Expert Syst Appl 214

Ryu JI et al (2021) Data-driven chemical kinetic reaction mechanism for F-24 jet fuel ignition. Fuel 290

Shang C, You F (2019) Data analytics and machine learning for smart process manufacturing: recent advances and perspectives in the big data era. Engineering 5(6):1010–1016

Shrestha KP, Seidel L, Zeuch T, Mauss F (2018) Detailed kinetic mechanism for the oxidation of ammonia including the formation and reduction of nitrogen oxides. Energy Fuels 32(10):10202–10217

Shrestha KP et al (2021) An experimental and modeling study of ammonia with enriched oxygen content and ammonia/hydrogen laminar flame speed at elevated pressure and temperature. Proc Combust Inst 38(2):2163–2174

Slavinskaya NA, Riedel U, Dworkin SB, Thomson MJ (2012) Detailed numerical modeling of PAH formation and growth in non-premixed ethylene and ethane flames. Combust Flame 159(3):979–995

Tian Z, Li Y, Zhang L, Glarborg P, Qi F (2009) An experimental and kinetic modeling study of premixed NH3/CH4/O2/Ar flames at low pressure. Combust Flame 156(7):1413–1426

United Nations Sustainable Development Goals. https://sdgs.un.org/goals

Verma S, Pant M, Snasel V (2021) A comprehensive review on NSGA-II for multi-objective combinatorial optimization problems. IEEE Access 9:57757–57791

Wakale AB, Banerjee S, Banerjee R (2018) Experimental and chemical kinetic study of the impact of n-butanol blending on the gross engine performance of a CRDI engine. Energy Convers Manag 178(2018):400–414

Wan K, Barnaud C, Vervisch L, Domingo P (2020) Chemistry reduction using machine learning trained from non-premixed micro-mixing modeling: application to DNS of a syngas turbulent oxy-flame with side-wall effects. Combust Flame 220:119–129

Chapter 7
Kinetic Mechanism Development and CFD Study of Combustion of Ammonia Blends in a Micro-gas Turbine

Dinesh Prabhakaran, Richie Shaji Mathew, Raja Banerjee, and Sayak Banerjee

Abstract Ammonia (NH_3) is a high potential zero-carbon fuel and renewable hydrogen carrier. However, its combustion properties are less favourable than hydrocarbons, with challenges such as lower flame speeds, longer ignition delay times, and a higher tendency for NO_x emissions. As a result, ammonia is typically blended with conventional fuels like methane, kerosene, and diesel for use in gas turbines or dual-fuel compression ignition (CI) engines, particularly during the early stages of implementation. This study focuses on two main objectives: (a) developing a semi-detailed $NH_3/CH_4/H_2/n\text{-}C_{12}H_{26}$ chemical mechanism involving 242 species and 1769 reactions, including sub-mechanisms for nitrogen oxides (NO_x) and soot formation, to analyse the combustion and emission behaviour of ammonia-fuel blends in engines; and (b) applying this mechanism to study the combustion properties of a simplified micro-gas turbine using a steady flamelet model. The validation of the mechanism was done using experimental and numerical data from the literature. The mechanism was integrated with CFD solver utilizing the steady flamelet approach to analyse the combustion characteristics of an 11 kW (thermal) micro-gas turbine combustor. Realizable $k-\varepsilon$ turbulence model was used in this work. Time averaged combustor exit temperature and emissions like NO_x, CO and unburnt fuel fractions were estimated. Though overall equivalence ratio was maintained to be lean, a rich-lean dual combustion regime was created with the help of dilution holes to control NO_x emissions. The effect of system pressure on emission characteristics was also investigated.

D. Prabhakaran · S. Banerjee
Department of Climate Change, Indian Institute of Technology Hyderabad, Sangareddy, Telengana, India

R. S. Mathew · R. Banerjee (✉) · S. Banerjee
Department of Mechanical and Aerospace Engineering, Indian Institute of Technology Hyderabad, Sangareddy, Telengana, India
e-mail: rajabanerjee@mae.iith.ac.in

© The Author(s), under exclusive license to Springer Nature Singapore Pte Ltd. 2026

H. Valera and A. K. Agarwal (eds.), *Ammonia and Low Carbon Combustion Engine Technologies for De-Fossilizing Transport Sector*, Energy, Environment, and Sustainability, https://doi.org/10.1007/978-981-96-6628-7_7

Keywords Ammonia combustion · Chemical kinetics · Flamelet model · Micro gas turbine

7.1 Introduction

Energy derived from combustion of fossil fuels has been the main stay of industrial revolution for the last two centuries. However, global warming and the associated climate change concerns due to greenhouse gas emissions from the combustion of such fossil fuels has led to global research initiatives into understanding combustion characteristics of newer fuels, particularly for carbon neutral fuels. Hydrogen and more recently ammonia stand out in this class of fuels. However, storage and transport of hydrogen is a major concern in several engineering applications like transportation. The gravimetric lower heat value of hydrogen is 120 MJ/kg, whereas its volumetric energy density reduces to about 2.1 MJ/l at 248 bar and subsequently increases to 8.5 MJ/l when liquified (Dimitriou and Javaid 2020). On the other hand, ammonia has a lower gravimetric heating value of 18 MJ/kg, and its volumetric energy density is 11.3 MJ/l. Therefore, ammonia is an excellent energy carrier of hydrogen. Moreover, ammonia is among the most commonly used industrial chemicals and therefore, extensive infrastructure for its production, storage and transportation exists globally (Aziz et al. 2020).

Though direct combustion of ammonia is not new (Verkamp et al. 1967), its utilization as fuel was very limited due to its poor combustion properties, like high ignition delay, very low laminar flame speed and high propensity to form NO_x due to fuel bound nitrogen (Kobayashi et al. 2019). Such properties may lead to flame instability, early quenching of flame and ammonia slip. However, there is renewed interest in direct combustion of ammonia as it is considered to be a carbon neutral fuel. Few good review papers have been published recently highlighting the fundamentals of ammonia combustion and its application to various combustor configurations like gas turbines, internal combustion engines and industrial furnaces (Dimitriou and Javaid 2020; Kobayashi et al. 2019; Elbaz et al. 2022; Herbinet et al. 2022).

Though direct ammonia combustion has been reported for several applications, the current review will only concentrate on gas turbine combustor configurations, keeping in mind the theme of the current work. Kobayashi and his co-workers have done some pioneering work to understand direct ammonia combustion in gas turbine applications. They have used laser diagnostics techniques such as Particle Image Velocimetry (PIV), and Planar Laser Induced Fluorescence (PLIF) imaging and Fourier Transform Infrared (FTIR) based gas analysis to study the flow field, flame structure and emissions characteristics in a micro gas turbine swirl combustor (Okafor et al. 2020). They studied these parameters for NH_3/CH_4 blends. They have also proposed a two-zone combustor design for non-premixed NH_3/air combustion where the global equation ratio was greater than 1 near the fuel injection zone which was subsequently diluted to much leaner conditions away from that through introduction of fresh air through dilution holes. This helped reduce NO_x emissions at the combustor exit (Somarathne et al. 2019, 2018). They have also numerically studied the combustion characteristics of premixed combustion for NH_3/air (Somarathne et al. 2017) and NH_3/CH_4 blends (Kovaleva et al. 2022) in a similar configuration. They reported that in both cases, NO and NH_3 slip was minimum at an equivalence ratio of approximately 1.2.

Numerical modelling has been used extensively to study combustion in gas turbines. Such modelling includes two important segments: (a) combustion reaction mechanism and (b) estimation of heat release and species composition in the flow domain. Some of the most well regarded detailed NH_3 combustion mechanisms are by Okafor et al. (2018), Li et al. (2019) and Shrestha et al. (2021). There are different approaches to integrate the reaction mechanism in a numerical Navier Stokes solver and simulate turbulent combustion. These include Eddy Dissipation Model (Magnussen and Hjertager 1977) and Shell/Characteristic Time Combustion (CTC) (Abraham et al. 1985) which assume infinitely fast reaction and contain only very few global reactions to simulate reactions. There are tuning parameters that can be used to have a correct estimate of the heat release rates. These are simple and computationally very inexpensive to implement in the numerical solver. On the other hand of the spectrum is the Finite Rate Model where several reactions can be coupled to the Navier Stokes solver. These are computationally very expensive as a large number of scalar transport species have to be tracked in the solver and even larger number of stiff ODEs are to be solved to track the creation and destruction of species due to the chemical reactions. Taken together, these calculations can become prohibitively very expensive and therefore, extreme reduction in combustion mechanisms are required which may lead to loss of fidelity in such mechanisms, particularly for larger fuel molecules like *iso*-octane, *n*-dodecane and others. Various methods have been explored in literature to accelerate chemistry calculations like in situ adaptive tabulation (ISAT) (Pope 1997), piecewise reusable implementation of solution mapping (PRISM) (Tonse et al. 2003), manifold-based methods like intrinsic low-dimensional manifolds (ILDM) (Maas and Pope 1992) and machine learning methods like ANN (Christo et al. 1996). The Eddy Dissipation Concept (EDC) couples finite rate chemistry and turbulence by assuming that a computational cell is a partially stirred reactor and there exist fine turbulent structures of certain fraction in the cells. Flamelet models are used when the combustion process is still fast and separation of reaction and turbulence scales are still possible. Therefore, scalar transport equations of one or multiple color functions are used to track the flame front. These flame fronts represent the transition from burnt to unburnt region in case of premixed combustion whereas it represents the transition from fuel side to the oxidizer side in non-premixed combustion. The color functions are often used to estimate composition in the flow domain with the help of pre-tabulated look-up tables. These are computationally less expensive and therefore can accommodate larger mechanisms.

Given below are few illustrative examples of numerical studies using the above methods with ammonia as a fuel in gas turbine applications. Sun et al. (2022) used the EDC method to simulate CH_4/NH_3 blend in an industrial gas turbine and they observed that increased NH_3 percentage in the blend led to the formation of longer and thinner lifted flame. Okafor et al. (2019) used a partially stirred reactor model in OpenFOAM with ISAT acceleration technique to simulate combustion in a micro gas turbine. Ilbas et al. (2021) reported co-firing of ammonia and kerosene in an industrial scale gas turbine burner. This studied the effect of swirl on flame location and effect of NH_3 blending on NO_x emissions. Bioche et al. (2021) studied the effect of NH_3/H_2 blending in a gas turbine burner. Their study was based on premixed combustion and using finite rate chemistry.

This work reports the construction of a semi-detailed $NH_3/CH_4/H_2$/n-$C_{12}H_{26}$ combustion mechanism consisting of 242 species and 1769 reactions and its validation. This mechanism was then used to simulate combustion of an 11 kW micro-gas turbine that was fired with blend of NH_3/CH_4 as fuel. Emission characteristics of NOx and NH_3 at the gas turbine exit were analyzed as function of gas turbine operating pressure and dilution air flowrate.

Transition to newer combustion technologies will involve co-firing of traditional fuels like diesel or natural gas with carbon neutral fuels like hydrogen and ammonia. Research in such dual fuel combustion systems has seen rapid progress over the last few years (Boretti 2019). Different combinations of dual fuel combustors have been investigated so far like diesel (liquid)—hydrogen (gas) (Bhagat et al. 2023), diesel (liquid)—methane (gas) (Misra et al. 2019) and ammonia (gas)—methane (gas) (Chai et al. 2021). Therefore, a combustion kinetics mechanism that can simulate combustion kinetics of all the major fuel components like diesel and natural gas surrogates along with hydrogen and its derivatives is an essential step in modelling the newer types of fuel blends. This work first reports the construction of an n-dodecane/methane/hydrogen/ammonia along with NOx and PAH submission mechanisms. This mechanism is then used to simulate combustion of CH_4/NH_3 blends in a realistic combustor of a micro-gas turbine and major combustion characteristics like flame structure, unburnt fuel and emissions are estimated.

7.2 Construction of Combustion Mechanism

Numerous comprehensive chemical reaction models for ammonia or ammonia/hydrocarbon mixtures have been developed recently. In 2018, Shrestha et al. (2018) developed a comprehensive combustion mechanism for H_2-CO-CH_4-NO-NH_3, utilizing extensive experimental data, including ignition delays, laminar flame speeds, and species concentrations, for validation. However, some NH_3-C_1 species and reactions were still missing in Shrestha's mechanism. To create a comprehensive mechanism for $NH_3/H_2/CH_4$, three detailed kinetic mechanisms for ammonia and H_2/CH_4—namely AramcoMech 2.0 (NUIG Combustion Chemistry Center n.d.), Shrestha and Tian et al. (2009) models—were combined in the work by Li et al. (2019). In 2021, Sreshtha et al. (2021) have updated their model to predict laminar flame speeds of NH_3/H_2 blends better. In the current work, Sreshtha's mechanism has been updated by their new one in the assembled mechanism of Li et al. Further, a skeletal n-dodecane mechanism by Luo et al. (2014) and polyaromatic hydrocarbon

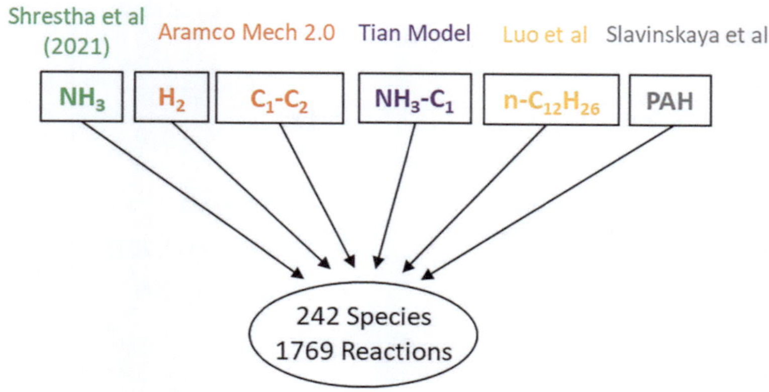

Fig. 7.1 Schematic of assembled $NH_3/CH_4/H_2$/n-$C_{12}H_{26}$ mechanism

chemistry set by Slavinskaya et al. (2012) have been added to the overall mechanism to arrive at a comprehensive H_2-CH_4-NH_3-n-$C_{12}H_{26}$ mechanism. Figure 7.1 shows the flow chart of the assembling process resulting in a mechanism of 242 species and 1769 reactions (MixDMAH242).

7.2.1 Validation Under Diffusion Flame Conditions

The kinetic model has been validated against experimental data available from open literature for a jet stirred reactor (JSR) and burner stabilized flame (BSF) setup. Figure 7.2 shows the comparison of MixDMAH242, Shreshtha model (Shrestha et al. 2021) and experimental data reported by Dayma and Dagaut (2006) for JSR setup, whereas the same combustion kinetics models were compared with the experimental data by Vandooren et al. (1994) for BSF setup. Dayma and Dagaut (2006) performed several experiments to study the oxidation of the hydrogen-oxygen-nitrogen based mixtures. The case with pressure of 10 atm and equivalence ratio of 0.5 in the presence of 200 ppm NO case was used for comparison in this validation. Vandooen et al. (1994) used molecular beam sampling mass spectrometry to determine the structure of a quasi-equimolar ammonia-nitric oxide flame over a BSF at 7.2 kPa. The initial composition was 47.2% of NO, 46.1% of NH_3 and 6.7% of Ar that makes the equivalence ratio of 1.46. The initial flow velocity was 0.6 m/s at 295 K. As can be seen from the figure, MixDMAH242 shows a good agreement with the experimental data. In all cases, MixDMAH242 follows the profile given by Shreshtha quite closely, other than the NO prediction in the JSR case. In this case instance, MixDMAH242 is under predicting NO profile.

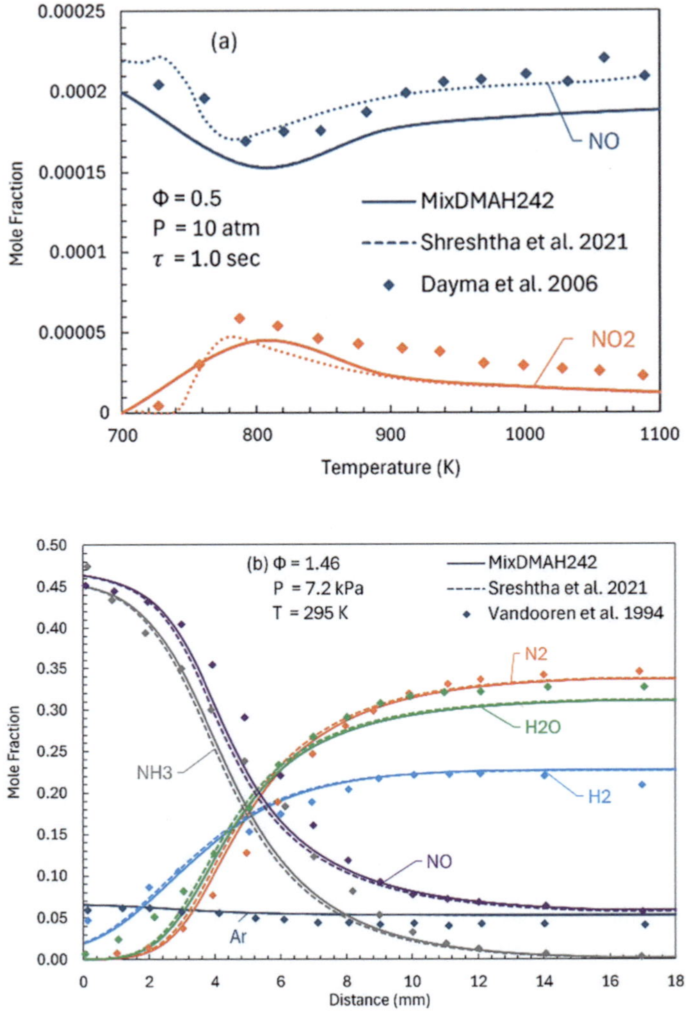

Fig. 7.2 Comparison between model prediction and experimental data for **a** H2/O2/NO (220 ppm)/ N2 φ = 0.5, 10 atm, τ = 1.0 s), JSR & **b** for NH3/NO/Ar (φ = *1.46), BSF*

7.3 Mathematical Formulation

7.3.1 Conservation Equations

Favre averaged Navier Stokes equation was used to simulate the fluid flow in the micro gas turbine used in this work. The continuity and momentum equation are of the form:

$$\frac{\partial \rho}{\partial t} + \dot{\nabla} \cdot (\rho \vec{u}) = 0 \tag{7.1}$$

$$\frac{\partial (\rho \vec{u})}{\partial t} + \nabla \cdot (\rho \vec{u} \vec{u}) = -\nabla p + \nabla \cdot \left[\mu_{eff} \left(\nabla \vec{u} + \nabla \vec{u}^T \right) \right] + S_m \tag{7.2}$$

The effective viscosity is the sum total of eddy viscosity, μ_t and fluid viscosity, μ. The source term, S_m is the body force due to gravity. The temperature field is obtained by solving the transport equation of total energy of the form:

$$\frac{\partial (\rho E)}{\partial t} + \nabla \cdot \left[\vec{u}(\rho E + p) \right] = \nabla \cdot \left[k_{eff} \nabla T - \sum_j h_j \vec{J}_j \right] + S_e \tag{7.3}$$

where, $E = h - \frac{p}{\rho} + \frac{u^2}{2}$, enthalpy of the jth species is given as $h_j = \int_{T_{ref}}^{T} c_{p,j} dT$ and the diffusion flux is determined as, $J_j = -\left(\rho D_{i,m} + \frac{\mu_t}{Sc_t} \right) \nabla y_i$. Turbulence is modelled using the $k - \varepsilon$ model and the eddy viscosity is obtained from the turbulent kinetic energy and its dissipation rate as, $\mu_t = \rho C_\mu \frac{k^2}{\varepsilon}$, where $C_\mu = 0.09$.

7.3.2 Combustion Model

Turbulent non-premixed combustion can be considered an ensemble of laminar flamelets. As combustion chemistry is assumed to be fast, separation of turbulent and combustion time scales can be assumed. Ensemble of multiple flamelets were calculated and stored in a pre-calculated lookup table and can be used to simulate non-premixed turbulent combustion. In non-premixed combustion, the flame front separates the fuel and the oxidizer sides, and mixture fraction, which is defined as $f = \frac{Z_i - Z_{i,ox}}{Z_{i,f} - Z_{i,ox}}$ can be used a color function to track the progress of combustion if the gradient normal to the flame front is significantly higher than that in the tangential direction. In such a case, a coordinate transformation from the physical space to the mixture fraction space can be done and the species mass fraction and the temperature distribution for a flamelet can be written as given in Peters (1988, 1999).

$$\rho \frac{\partial y_i}{\partial t} = \frac{1}{2} \rho \chi \frac{\partial^2 y_i}{\partial f^2} + S_i \tag{7.4}$$

$$\rho \frac{\partial T}{\partial t} = \frac{1}{2} \rho \chi \frac{\partial^2 T}{\partial f^2} - \frac{1}{C_p} \sum_i h_i S_i + \frac{1}{2 C_p} \rho \chi \left[\frac{\partial C_p}{\partial f} + \sum_i C_{p,i} \frac{\partial y_i}{\partial f} \right] \frac{\partial T}{\partial f} \tag{7.5}$$

where S_i is the $i\,th$ species reaction rate and h_i is the specific enthalpy of the ith species. The basic Naviers Stokes solver is coupled with two transport equation of the form:

$$\frac{\partial(\overline{\rho}\tilde{f})}{\partial t} + \nabla \cdot (\overline{\rho}\tilde{u}\tilde{f}) = \nabla \cdot \left[\left(\frac{\tilde{k}}{C_p} + \frac{\mu_t}{\sigma_t}\right)\nabla\tilde{f}\right] + S_m \tag{7.6}$$

And

$$\frac{\partial}{\partial t}\left(\overline{\rho}\widetilde{f''^2}\right) + \nabla \cdot \left(\overline{\rho}\tilde{\mathbf{u}}\widetilde{f''^2}\right) = \nabla \cdot \left[\left(\frac{\tilde{k}}{C_p} + \frac{\mu_t}{\sigma_t}\right)\nabla\widetilde{f''^2}\right] + C_g\mu_t \cdot (\nabla\tilde{f}) - C_d\overline{\rho}\frac{\tilde{\varepsilon}}{\tilde{k}}\widetilde{f''^2} \tag{7.7}$$

The first equation is the transport of the mean value of mixture fraction and the second equation is the transport of its second moment. The second moment of mixture fraction represents the expected variation in the instantaneous values of mixture fraction and this is represented by an assumed probability distribution function, which is a β function in this case. Mean mass fraction can be calculated from,

$$\widetilde{\phi_i} = \int_0^1 p_f(f)p_\chi(\chi_{st})\phi_i(f, \chi_{st})dfd\chi_{st} \tag{7.8}$$

$p_\chi(\chi_{st})$ is a delta function of the form, $p_\chi = \delta(\chi - \overline{\chi})$ and $\chi_{st}$is estimated as, $\overline{\chi_{st}} = \frac{C_\chi\tilde{\varepsilon}\widetilde{f''^2}}{\tilde{k}}$. Therefore, all the species mass fractions can be tabulated apriori as a function of $\tilde{f}$ and $\widetilde{f''^2}$.

7.4 Results and Discussion

The current study was performed in a micro-gas turbine (MGT). It consists of a cylindrical main chamber of 80 mm diameter and 100 mm in height. There is a conical exit where the diameter reduces to 40 mm. The fuel is introduced from a centrally located orifice of 3.6 mm and co-flowing air is introduced from an annular section of internal diameter of 16.10 mm and outer diameter of 21.47 mm. There are 10 holes of 10 mm diameter each at a height of 40 mm from the bottom. The fuel flowrates were maintained so that the input energy was 11 kW. In all the cases simulated here, 70% energy came from ammonia and the remaining 30% came from methane. The geometry of the MGT is shown in Fig. 7.3. The domain is discretized with a mesh of 1.77 million cells. Commercial computational fluid dynamics solver Ansys Fluent v22 R2 was used for all the simulations reported here.

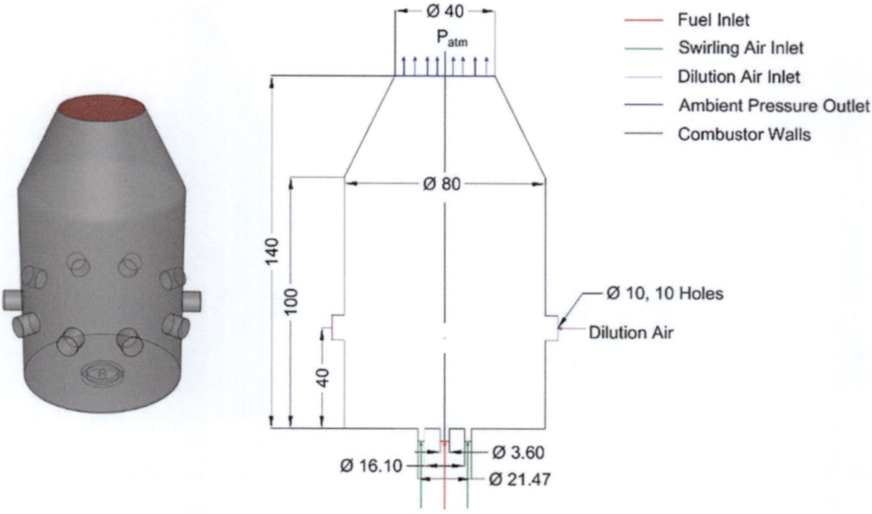

Fig. 7.3 Schematic of the micro-gas turbine

7.5 Base Case

Here the system pressure was maintained at 10 bar pressure. The fuel was injected at a velocity of 6.92 m/s. The axial velocity of the co-flowing air was 1.64 m/s and the swirl number was taken as 0.5. Therefore, the tangential component of co-flowing air was 0.82 m/s. Based on these flow conditions, the equivalence ratio at the combustor inlet was maintained at 1.2. This was done to starve NH_3 of oxygen and prevent the formation of fuel driven NO_x. It is expected that formation of H_2 will be favored due to thermal breakdown of NH_3. No dilution air was introduced in this case and therefore the overall equivalence ratio was also maintained at 1.2.

Figure 7.4 shows the contour plots of axial velocity and temperature along the center plane of the combustor. A small recirculation zone is seen in the annular space between the fuel injector and co-flowing air near the entrance due to the bluff body effect of the burner. The swirling air forms a waist that helps form the recirculation zone. This helps anchor the flame near the combustor entrance. The flame grows with axial distance due to mixing of air and fuel streams. Figure 7.5 shows the NO_x and ammonia mass fraction profiles along the center plane. Here NO_x is the sum total of the mass fractions of NO, NO_2 and N_2O. It follows the temperature closely. It is a thin section near the entrance and becomes diffused near the exit. The concentration of NH_3 is high near the fuel entrance and monotonically decreases away from the entrance. As seen from the figure, there is still some unburnt ammonia at the exit. The emission at the exit of the combustion is 8.51×10^{-5} g/kJ for NO_x, 1.62×10^{-2} g/kJ for NH_3 and 1.54×10^{-3} kJ/kJ for CO.

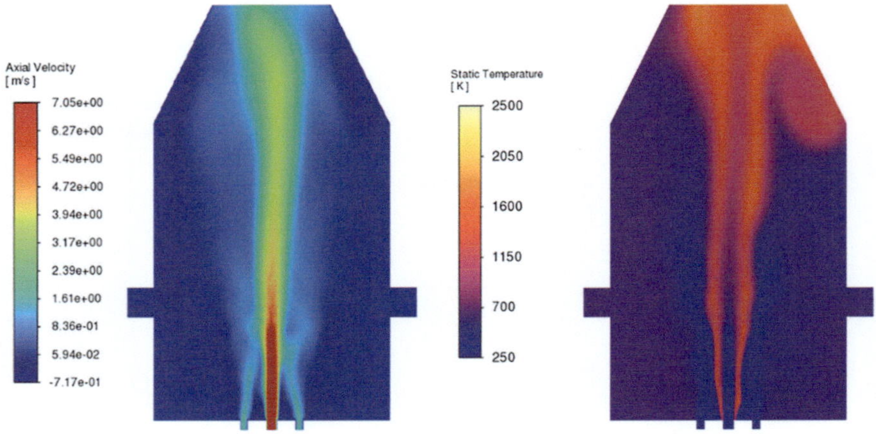

Fig. 7.4 Axial velocity (left) and temperature contours (right) along the center plane of the combustor at 10 bar

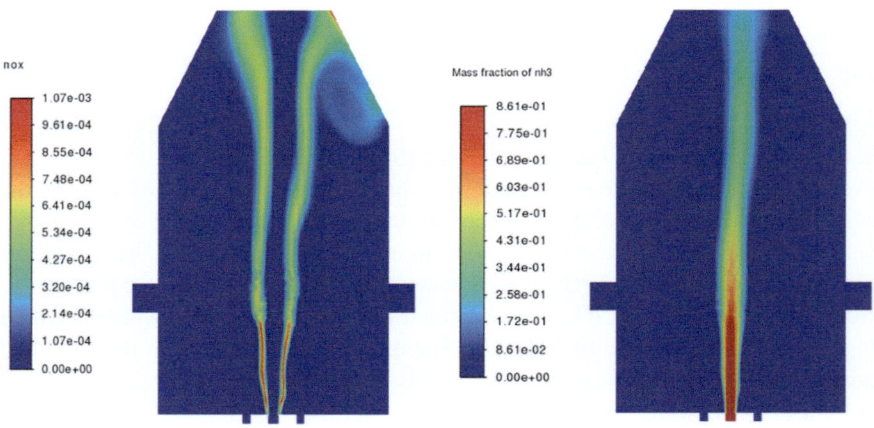

Fig. 7.5 Contours of Nitrogen Oxides (NOx) (left), Ammonia (right) along the center plane at 10 bar

7.6 Parametric Study

As can be seen from the base case analysis, unburnt ammonia is seen escaping from the combustor. Moreover, there are few reactions in the combustion mechanism that are pressure sensitive. These are:

$$NH_3 + OH \rightarrow NH_2 + H_2$$

$$NH_2 + NO \rightarrow N_2 + H_2O$$

$$NH_2 + NO \rightarrow NNH + OH$$

$$NH_2 + HO_2 \rightarrow H_2NO + OH$$

At elevated pressure the forward reactions are favored. Reaction 4 produces OH radical which triggers the consumption of NH_3 and produced NH_2 radical, which in turn further triggers the consumption of NO. Taken together, it is expected that both ammonia and NO emissions should decrease at elevated pressures. Furthermore, dilution air should help oxidize the thermally cracked H_2 produced in the rich region of the combustor. Keeping these in mind, a parametric study is conducted to see the effect of dilution air and system pressure on the emission at the exit. The dilution air flowrate was varied in a manner such that the global equivalence ratio is varied from 0.5–1.2. 1 and 10 bar system pressures were analyzed.

Air flowrate from the dilution holes increases with decrease in the global equivalence ratio. The equivalence ratio at the fuel inlet side is kept constant at 1.2. As described before, this was done to starve ammonia of oxygen that will prevent formation of NO_x due to fuel bound nitrogen. However, the hydrogen thus formed would then be oxidized by the supply of fresh air entering the combustor through the dilution holes. This is the so called two zone combustor proposed by Somarathne et al. (2019). Figure 7.6 shows the flame shape with the increase in global equivalence ratio and system pressure. As can be seen from the figure, there is not much change in the shape of the flame with dilution, even though the flame becomes marginally narrower at $\phi_{global} = 0.5$ as compared to $\phi_{global} = 1.2$. However, the flame shape becomes narrower more prominently for the 10 bar case under leaner conditions.

Figure 7.7 shows the area weighted average of various scalars at the exit plane of the combustor. Area weighted average of scalar φ is defined as,

$\varphi_{avg} = \frac{1}{A}\int_A \varphi dA.$

Exit temperature decreases monotonically as the mixture becomes lean for both the pressure cases, with temperatures at 10 bar being slightly higher than at 1 bar. This decrease in temperature at leaner conditions is expected because a larger amount fresh air is entering the combustor at leaner conditions leading to a sensible cooling of the combustion gases.

Unburnt methane decreases with decrease in ϕ_{global} and increase in system pressure. Such a decrease is expected because leaner mixture increases the propensity of complete combustion and higher pressures also favor the forward reactions. Likewise, unburnt ammonia also decreases under leaner conditions and higher pressures. The pressure sensitive reactions have already been presented before. Unburnt hydrogen increases with pressure while it decreases under leaner conditions. As with the other unburnt fuel fractions, higher availability of oxygen under leaner conditions favoring a more complete oxidation is also valid here. However, the higher amount of hydrogen in the exhaust is due to the higher propensity of the NH_3 breaking down into NH_2 and H_2 at elevated pressure. NO_x is marginally increasing under leaner conditions. This may be because some of the fuel bound nitrogen is reacting with the excess

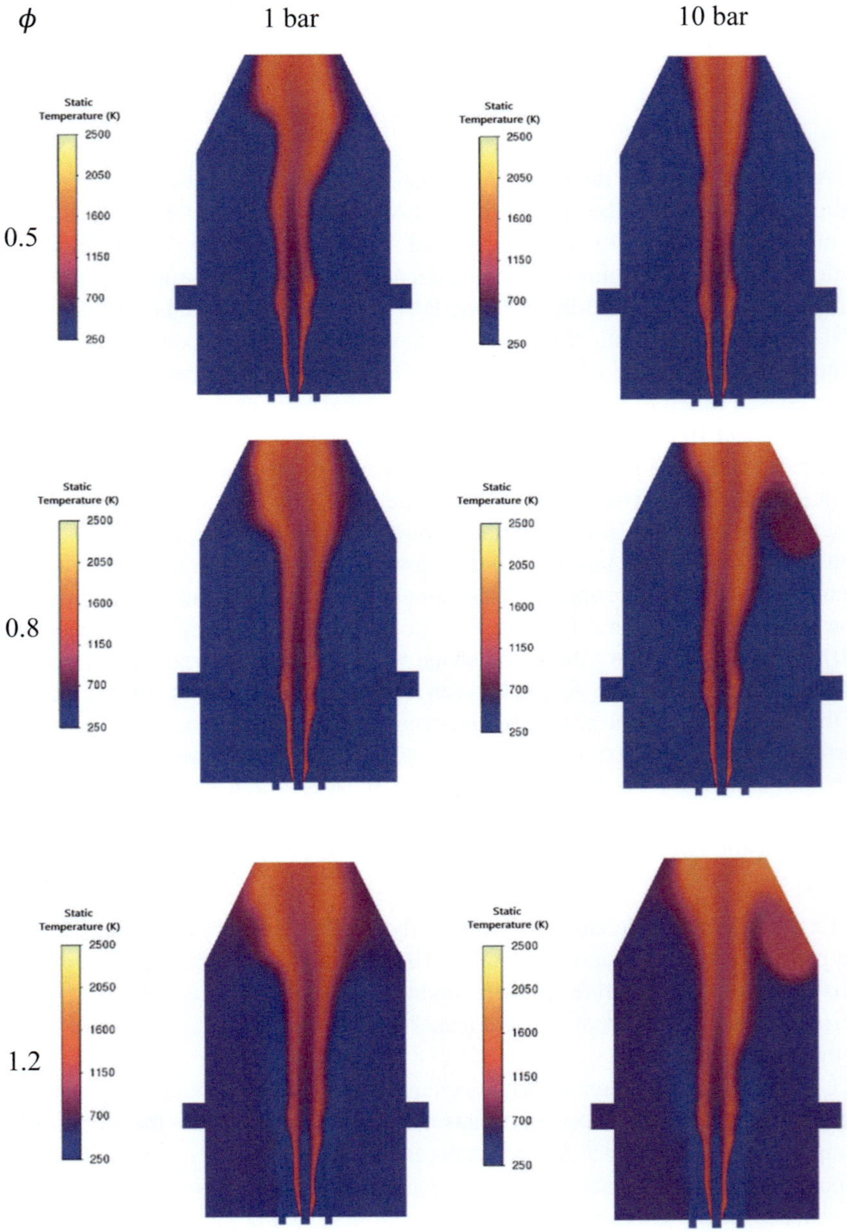

Fig. 7.6 Mid-plane temperature contours plots for $\phi_{\text{global}} = 0.5$, 0.8 and 1.2 at 1 and 10 bars

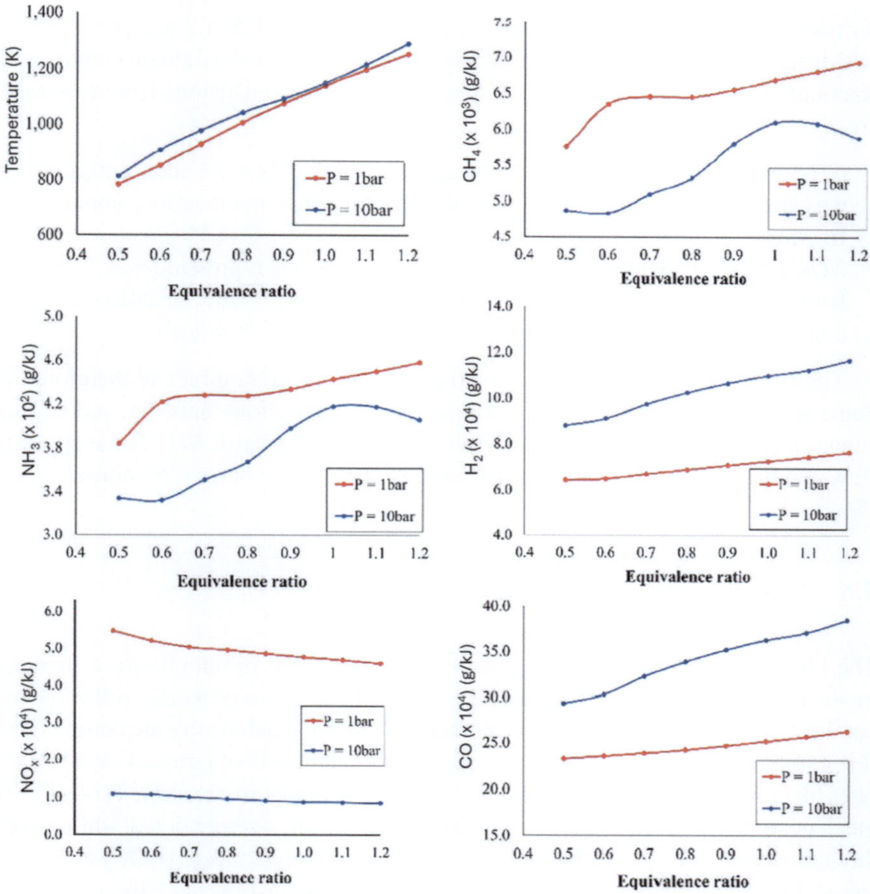

Fig. 7.7 Variation in temperature and mass fractions of unburnt fuel, hydrogen, NOx and CO at the exit plane of the combustor with equivalence ratio

oxygen available under leaner conditions. However, there is a significant decrease in NO_x emission at elevated pressure. The four pressure sensitive reactions presented before are therefore playing an important role in the overall NO_x emission. *CO* emissions are also decreasing with the availability of excess air. However, *CO* emissions increase with pressure.

7.7 Conclusion

In this work, the construction of a 242 species and 1769 reaction combustion mechanism along with its validation to predict NO_x and NH_3 profiles was presented. This mechanism was then used to simulate turbulent combustion in a 11-kW micro gas turbine combustor. A global equivalence ratio sweep from 0.5–1.2 was performed at two different system pressures to examine the effects of dilution ratio on unburnt fuel fractions, *NOx* and *CO* emissions. Some of the major conclusions from this work are:

- NH_3 emissions was two orders of magnitude higher than all other emissions. It was also found that leaner combustion is generally favored to reduce unburnt fuel fractions.
- *NOx* emissions reduced by a factor of almost six when pressure was increased from 1 to 10 bar. This is because there is a set of four pressure sensitive reactions that favor utilization of *NO* and favor the production of hydrogen.

The current work was done on co-axial injection configuration and therefore the flame structure was long. Other fuel injection configurations must be studied that enhance the mixing of air and fuel. This is important because NH_3 flame speed is slow and therefore enhanced mixing should help reduce unburnt NH_3 emissions.

7.8 Limitations of the Study

The kinetic mechanism, that has been used, is large and is suitable for a flamelet model based turbulent combustion analysis. The flamelet model works well when the combustion kinetics are fast and the thermal and mass diffusion rates are comparable. It is computationally inexpensive and gives accurate results in terms of heat release rates but tends to be inconsistent when the chemical reactions are slow. Typical *NOx* mechanism are slow and therefore, *NOx* emissions are overpredicted while using the flamelet model. Other options like Eddy Dissipation Concept (EDC) are better suited for *NOx* estimation. However, EDC based combustion modelling with this mechanism will be prohibitively expensive as this mechanism consists of a large number of stiff reactions.

7.9 Future Scope

The current simulations were performed using RANS based turbulence models. However, it is well reported in literature that LES based turbulence gives improved combustion prediction as it resolves turbulence up to the inertial scales. Another improvement that needs to be addressed is to include heat transfer due to radiation because carbon neutral fuels produce a significant amount of water vapor as combustion products which makes the combustion zone optically thick. The combustion modelling performance is expected to improve significantly if these two modelling paradigms are included.

References

Abraham J, Bracco FV, Reitz RD (1985) Comparisons of computed and measured premixed charge engine combustion. Combust Flame 60(3):309–322

Aziz M, TriWijayanta A, Nandiyanto ABD (2020) Ammonia as effective hydrogen storage: a review on production, storage and utilization. Energies (Basel) 13(12)

Bhagat RN et al (2023) A review of performance and emissions of diesel engine operating on dual fuel mode with hydrogen as gaseous fuel. Int J Hydrog Energy 48(70):27394–27407

Bioche K, Bricteux L, Bertolino A, Parente A, Blondeau J (2021) Large eddy simulation of rich ammonia/hydrogen/air combustion in a gas turbine burner. Int J Hydrogen Energy 46(79):39548–39562

Boretti A (2019) Advantages and disadvantages of diesel single and dual-fuel engines. Front Mech Eng 5:64

Chai WS et al (2021) A review on ammonia, ammonia-hydrogen and ammonia-methane fuels. Renew Sustain Energy Rev 147:111254

Christo FC, Masri AR, Nebot EM (1996) Artificial neural network implementation of chemistry with pdf simulation of H_2/CO_2 flames. Combust Flame 106(4):406–427

Dayma G, Dagaut P (2006) Effects of air contamination on the combustion of hydrogen-effect of NO and NO_2 addition on hydrogen ignition and oxidation kinetics. In: combustion science and technology, pp 1999–2024

Dimitriou P, Javaid R (2020) A review of ammonia as a compression ignition engine fuel. Int J Hydrogen Energy 45(11):7098–7118

Elbaz AM, Wang S, Guiberti TF, Roberts WL (2022) Review on the recent advances on ammonia combustion from the fundamentals to the applications. Fuel Commun 10:100053

Herbinet O, Bartocci P, Grinberg Dana A (2022) On the use of ammonia as a fuel—a perspective. Fuel Commun 11:100064

Ilbas M, Kumuk O, Karyeyen S (2021) Numerical study of a swirl gas turbine combustor for turbulent air and oxy-combustion of ammonia/kerosene fuels. Fuel 304

Kobayashi H, Hayakawa A, Somarathne KDKA, Okafor EC (2019) Science and technology of ammonia combustion. Proc Combust Inst 37(1):109–133

Kovaleva M et al (2022) Numerical and experimental study of product gas characteristics in premixed ammonia/methane/air laminar flames stabilised in a stagnation flow. Fuel Commun 10:100054

Li A, Konnov A, He G, Qin F, Zhang D (2019) Chemical mechanism development and reduction for combustion of NH3/H2/CH4 mixtures. Fuel 257

Luo Z et al (2014) Development and validation of an n-dodecane skeletal mechanism for spray combustion applications. Combust Theor Model 18(2):187–203

Maas U, Pope SB (1992) Simplifying chemical kinetics: intrinsic low-dimensional manifolds in composition space. Combust Flame 88(3–4):239–264

Magnussen BF, Hjertager BH (1977) On mathematical modeling of turbulent combustion with special emphasis on soot formation and combustion. Symp (Int) Combust 16(1):719–729

Misra A et al (2020) Methane–diesel dual fuel engine: a comprehensive review. In: Proceedings of international conference in mechanical and energy technology: ICMET 2019, India. Springer Singapore

NUIG Combustion Chemistry Center. https://c3.universityofgalway.ie/combustionchemistrycentre/mechanismdownloads/

Okafor EC et al (2018) Experimental and numerical study of the laminar burning velocity of CH4–NH3–air premixed flames. Combust Flame 187:185–198

Okafor EC et al (2019) Towards the development of an efficient low-NOx ammonia combustor for a micro gas turbine. Proc Combust Inst 37(4):4597–4606

Okafor EC et al (2020) Control of NOx and other emissions in micro gas turbine combustors fuelled with mixtures of methane and ammonia. Combust Flame 211:406–416

Peters N (1988) Laminar flamelet concepts in turbulent combustion. Symp (Int) Combust 21(1). Elsevier

Peters N (1999) The turbulent burning velocity for large-scale and small-scale turbulence. J Fluid Mech 384:107–132

Pope SB (1997) Computationally efficient implementation of combustion chemistry using in situ adaptive tabulation. Combust Theor Model 1(1):41–63

Shrestha KP, Seidel L, Zeuch T, Mauss F (2018) Detailed kinetic mechanism for the oxidation of ammonia including the formation and reduction of nitrogen oxides. Energy Fuels 32(10):10202–10217

Shrestha KP et al (2021) An experimental and modeling study of ammonia with enriched oxygen content and ammonia/hydrogen laminar flame speed at elevated pressure and temperature. Proc Combust Inst 38(2):2163–2174

Slavinskaya NA, Riedel U, Dworkin SB, Thomson MJ (2012) Detailed numerical modeling of PAH formation and growth in non-premixed ethylene and ethane flames. Combust Flame 159(3):979–995

Somarathne KDKA, Hatakeyama S, Hayakawa A, Kobayashi H (2017) Numerical study of a low emission gas turbine like combustor for turbulent ammonia/air premixed swirl flames with a secondary air injection at high pressure. Int J Hydrogen Energy 42(44):27388–27399

Somarathne KDKA, Colson S, Hayakawa A, Kobayashi H (2018) Modelling of ammonia/air non-premixed turbulent swirling flames in a gas turbine-like combustor at various pressures. Combust Theor Model 22(5):973–997

Somarathne KDKA et al (2019) Emission characteristics of turbulent non-premixed ammonia/air and methane/air swirl flames through a rich-lean combustor under various wall thermal boundary conditions at high pressure. Combust Flame 210:247–261

Sun J, Yang Q, Zhao N, Chen M, Zheng H (2022) Numerically study of CH4/NH3 combustion characteristics in an industrial gas turbine combustor based on a reduced mechanism. Fuel 327

Tian Z, Li Y, Zhang L, Glarborg P, Qi F (2009) An experimental and kinetic modeling study of premixed NH3/CH4/O2/Ar flames at low pressure. Combust Flame 156(7):1413–1426

Tonse SR, Moriarty NW, Frenklach M, Brown NI (2003) Computational economy improvements in PRISM. Int J Chem Kinet 35(9):438–452

Vandooren J, Bian J, Van Tiggelen PJ (1994) Comparison of experimental and calculated structures of an ammonianitric oxide flame. Importance of the NH2 + NO reaction. Combust Flame 98(4):402–410

Verkamp FJ, Hardin MC, Williams JR (1967) Ammonia combustion properties and performance in gas-turbine burners. Symp (Int) Combust 985–992

Part III
Ammonia Engine Technologies

Chapter 8
State-Of-The-Art Review on Ammonia-Based Diesel Engines

Sridhar Sahoo, Akash Kumar, P. Jayraj, and Dhananjay Kumar Srivastava

Abstract The chapter reviews the potential of ammonia as a fuel for diesel engines in the transportation sector, focusing on its advantages, challenges, and technological adaptations. Ammonia, with its carbon-free composition, offers significant environmental benefits by reducing greenhouse gas emissions. The study explores ammonia's combustion characteristics, energy density, and storage requirements. Technological advancements, such as dual-fuel systems and catalytic converters, are also evaluated for optimizing ammonia usage in diesel engines. The review also addresses economic and safety considerations, highlighting the feasibility of ammonia as a sustainable alternative to conventional diesel fuel in achieving cleaner transportation solutions.

Keywords Ammonia · Diesel engine · Fuel

8.1 Introduction

The demand for sustainable energy approaches has increased in recent years due to the increasing global population and economic demands. The existing conventional fossil fuels deplete fast and produce harmful greenhouse gases, affecting the ozone layer and human health. Diesel consumption is 70, 13, and 17% by transportation, agriculture, and other industrial applications (Press Information Bureau 2022). The existing diesel engines used in the transportation and agriculture emit harmful gases such as CO, CO_2, HC, NOx, particulates, and smoke due to their poor combustion and presence of long-chain hydrocarbon compounds. The emissions were reduced previously through the optimum design of the combustion chamber and implementation of after-treatment technologies such as selective catalytic reduction (SCR),

S. Sahoo (✉) · A. Kumar · P. Jayraj
Agricultural and Food Engineering Department, IIT Kharagpur, Kharagpur, India
e-mail: sridharsahoo366@agfe.iitkgp.ac.in

D. K. Srivastava
Department of Mechanical Engineering, IIT Kharagpur, Kharagpur, India

© The Author(s), under exclusive license to Springer Nature Singapore Pte Ltd. 2026
H. Valera and A. K. Agarwal (eds.), *Ammonia and Low Carbon Combustion Engine Technologies for De-Fossilizing Transport Sector*, Energy, Environment, and Sustainability, https://doi.org/10.1007/978-981-96-6628-7_8

diesel particulate filter (DPF), and diesel oxidation catalyst (DOC). However, the CO and NOx persist due to carbon in the diesel. The IPCC has highlighted that rising greenhouse gas emissions contribute to the increase in global temperature (Change et al. 2014). The carbon pricing mechanism and the Paris Agreements have been implemented to limit the rise of global temperature and achieve net-zero emissions (Tvinnereim and Mehling 2018). In this context, alternative fuels for diesel engines are gaining popularity due to their potential emission advantages over conventional diesel fuel. Alternative fuels such as biodiesel, hydrogen, compressed natural gas, dimethyl ether, alcohol fuel, and ammonia are promising in complete or partial substitution of diesel fuel from the existing diesel engines.

Ammonia (NH_3), a chemical compound consists of nitrogen and hydrogen, is a potential alternative fuel for diesel engines. The major benefit of using ammonia lies in its high hydrogen and zero-carbon content, which minimizes the formation of carbon emissions upon combustion. In contrast to hydrogen fuel, it can be easily stored and more conveniently transported, adding a real-world feasibility scenario. In addition, the production methodology and manufacturing infrastructure are available to integrate it into the existing energy grid. However, some limitations must be addressed before harnessing the fuel in the existing system. It has lower energy density, flammability limits, higher auto-ignition temperature, and NOx emissions, which require substantial research and optimization to develop and apply ammonia in the existing diesel engine (Kurien and Mittal 2022).

The chapter initially overviews the fundamental physicochemical properties of ammonia and its feasibility as an alternative fuel. Further, a brief review of ammonia combustion characteristics in diesel engines and diesel-like environments is outlined, considering the modifications and related combustion technologies required to utilize the fuel effectively. The subsequent section quantifies the overall performance and emissions to observe the challenges associated with future ammonia engines. Lastly, an overview of the cost and life cycle assessment of ammonia use is emphasized. The chapter closes with the future trends and open ends that should be researched. This review aims to give researchers and industry persons a brief understanding of the current trends and future opportunities for developing ammonia-based diesel engines.

8.2 Fundamentals of Ammonia as a Fuel

8.2.1 Physicochemical Properties of Ammonia

Ammonia (NH_3) comprises three hydrogen atoms and one nitrogen atom in a trigonal pyramid structure. Since it doesn't have carbon at the molecular level, it is 100% carbon-free, making it an attractive candidate for clean energy applications. It is an alkaline and colorless compound in nature, having a sharp, pungent smell. It has a molecular weight of 17.03 g mole^{-1}, and boiling and freezing points of 240 K and

Table 8.1 Physicochemical properties of ammonia (Dimitriou and Javaid 2020; Li et al. 2021)

Property	Ammonia (NH_3)	Diesel ($C_nH_{1.8n}$)
Hydrogen content	17.65% by mass	14% by mass
Carbon content	82.35% by mass	86% by mass
Stoichiometric air–fuel ratio	6.04	14.5
Molecular weight (M)	17.031 g mol^{-1}	170
Critical temperature (T_{cr})	405.15 K	720 K
Critical pressure (P_{cr})	111.5 atm	20 bar
Density at STP (ρ)	0.769 kg m^{-3}	837 kg m^{-3}
Lower heating value (LHV)	18.8 MJ kg^{-1}	42.5 MJ kg^{-1}
Autoignition temperature (T_{ig})	924 K	483 K
Flammability limit (vol. of air)	16–25%	0.6–5.5%

195.5 K, respectively. Under atmospheric conditions, its density is 0.769 kg/m^3, and its auto-ignition temperature is 924 K. The physicochemical properties of ammonia are detailed in Table 8.1.

Ammonia is primarily used as fertilizer in the agricultural industry. Apart from that, manufacturing units (textiles, plastics, dyes, and explosives), refrigeration plants, and water purification units also use ammonia in a small proportion. Gaseous ammonia has a strong, pungent smell and is easily detected by the human nose at a low concentration below 10 ppm. A high ammonia concentration is poisonous and thus has a threshold concentration limit for the human body. The "Immediately Dangerous to Life or Health (IDLH)" concentration for ammonia is 300 ppm (Nur Syamimi 2018). Due to its low density compared to air, it quickly dissipates into the upper region of the atmosphere upon accidental leakage, thereby reducing the risk. Further, ammonia detectors are available with sensors that can detect leaks at different concentration levels. Ammonia has a high volumetric energy density in liquid form due to its superior physicochemical properties than liquid hydrogen and other conventional liquid fuels. Ammonia has hydrogen atoms, thus acting as a hydrogen energy carrier and used as a standalone fuel in fuel cell applications.

8.2.2 *Production and Storage Infrastructure for Ammonia*

In 1774, ammonia was first isolated by Joseph Priestley (McKIE 1933), and almost 100 years later, the nitrogen fixation process was developed, where atmospheric nitrogen is converted to ammonia by the nitrogenase enzyme coupled with hydrolysis and hydrogen formation. The higher cost associated with the process and the nitrogenase enzyme production compatibility in the presence of atmospheric oxygen reduced its commercialization (Travis and Travis 2018). In 1908, Fritz Haber developed an ammonia synthesis process from its constituent elements and sold the patent to BASF Chemicals for ammonia production at a commercial scale. The Haber–Bosch process synthesizes ammonia by combining three nitrogen molecules from the air with hydrogen molecules under a pressure of 100–250 bar and temperature of 723–873 K in the presence of an iron catalyst. The catalyst helps break the triple bond present in the nitrogen molecule and adsorbs to the catalyst surface (Haber 2002). Subsequently, hydrogen adsorption takes place, forming ammonia. The nitrogen used in the process is produced using a well-established air separation unit that forms pure nitrogen from atmospheric air at a high efficiency and reasonable cost. The hydrogen used in ammonia synthesis is produced from various sources such as steam methane reforming, electrolyzers, etc., detailed in Pattabathula and Richardson (2016). The steam methane reforming process releases 9 to 10 tons of equivalent carbon dioxide per ton of hydrogen production, taking the process under the high-carbon footprint radar. The electrolyzer consumes electricity to separate hydrogen and oxygen molecules from the water, ensuring a zero-carbon footprint. However, the electric power should be renewable to ensure the process is carbon-free. The cost associated with electrolyzers is too high for widespread commercialization. Recently, a solid-state electrochemical process was developed where steam was decomposed at an anode and formed a hydrogen atom. The hydrogen atoms adsorb and are stripped of electrons, which are conducted to the cathode through the proton-conducting electrolytes. The hydrogen atom regains electrons at the anode and adsorbs with nitrogen atoms to form ammonia. Figure 8.1 shows the schematic layout of the process involved in ammonia production (Kurien and Mittal 2022).

Ammonia storage is a critical aspect after manufacturing, requiring special attention in designing the infrastructure, ensuring efficient handling, and minimizing sustainable environmental effects. Ammonia can be stored in liquid and gaseous

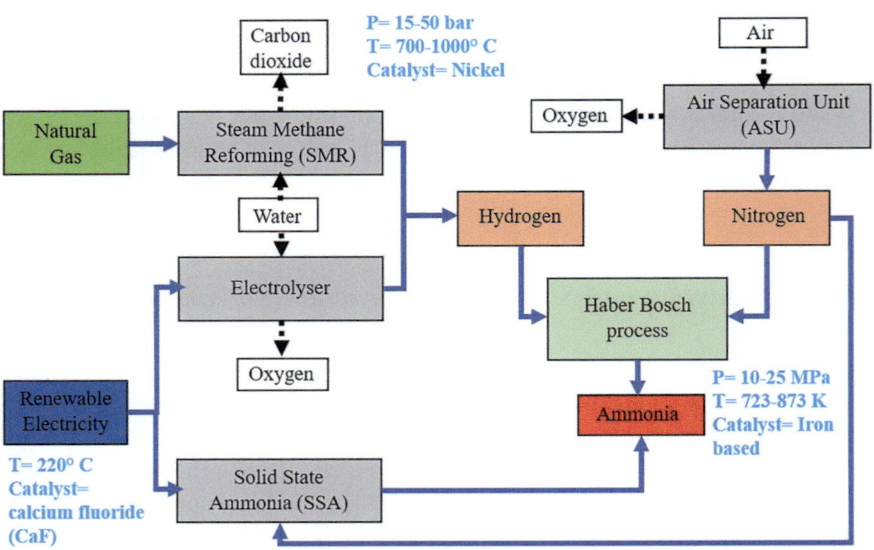

Fig. 8.1 Schematic layout of the ammonia production process (Kurien and Mittal 2022)

forms depending on the need and availability of facilities. Ammonia can be stored in liquid form in specially designed refrigerated tanks, ensuring no heat loss at low temperatures (240°K) or ambient pressure. Ammonia in liquid form can be stored in large tanks for long-term requirements and transported for long-distance requirements (Ishaq and Crawford 2024). Ammonia can be stored in pressurized storage tanks at a relatively low pressure of 10–15 bar, ambient temperature, for short-term storage and short-distance requirements. In recent years, solid storage technologies have offered ammonia's integration into solid materials (zeolite and activated carbon), like salts or porous media, to reduce safety concerns and improve ammonia transportation in conventional mode (Yang et al. 2023). Facilities such as leak detection sensors, automatic shut-off valves, and alarm systems must be employed in storage infrastructure to reduce possible hazards. The storage area must be adequately ventilated and integrated with a scrubbing system to contain the ammonia leakage and minimize the emissions. The storage facilities and related accessories must adhere to safety standards, considering ammonia's toxic and corrosive nature. Ammonia transportation depends on its supply chain, quantity, geographical locations, and distance. The pipelines offer a conventional method of continuous transportation for large volumes of ammonia over long distances with low risk. Road or rail transport is a suitable mode of transport for shorter distances. However, safety features need to be addressed to prevent spillage, leakage, and health hazards. Marine vessels with specialized refrigerated tanks are utilized for international transport between countries.

8.3 Ammonia Combustion in Diesel Engines

The 1940s marked a pivotal period for developing ammonia-fueled internal combustion engines. Amid the widespread destruction of oil refineries and storage depots during World War II, ammonia emerged as a viable and accessible alternative to power armored fighting vehicles and other military machinery. In 1943, liquid anhydrous ammonia was first successfully used as a fuel for motor buses, showcasing its potential beyond conventional fuels. Kroch (1945) published a seminal paper highlighting the promising results of burning ammonia mixed with coal gas, noting that this combination maintained engine power without causing corrosion or increasing lubricating oil consumption, unlike traditional gas oil engines. In his conclusion, Kroch emphasized that the coal gas, used as an ignition promoter, could easily be replaced by other gases like hydrogen, paving the way for a fully coal-independent operation, further expanding the potential of ammonia as a fuel. Ammonia (NH_3) has a low burning velocity and energy output compared to other fuels, such as methane. Its maximum laminar burning velocity (S_L) is significantly lower than methane, with ammonia flames showing a peak S_L of 7 cm/s versus methane's 35 cm/s. Ammonia also has a narrow flammability range (0.63–1.4) and a high auto-ignition temperature, complicating its fuel use. Increasing the oxygen concentration in ammonia/air mixtures can raise the flame velocity, making it closer to that of methane flames.

The flame speed generally decreases with higher pressure and slightly increases with equivalence ratio but becomes less variable at higher ratios. Regarding emissions, ammonia combustion produces less NO at lean conditions than methane but can produce higher NOx emissions under rich conditions. The mole fraction of NO in ammonia/air flames peaks at an equivalence ratio of 0.9 and decreases rapidly with higher ratios. Rich combustion conditions can reduce NOx but result in higher unburned ammonia. A slightly rich fuel condition around an equivalence ratio of 1.1 is recommended to optimize ammonia combustion, as it balances improved combustion efficiency with reduced NOx emissions (Dong et al. 2023).

The combustion of ammonia in dual-fuel operation is a realistic conception as the secondary fuel, with a lower auto-ignition temperature fuel can be used to trigger the combustion of the air-fuel mixture. Diesel was one of the first fuels tested for ignition in a dual-fuel engine in 1966. The researchers managed to operate a diesel-ammonia dual-fuel compression ignition engine with a compression ratio as low as 15.2:1 compared to the 35:1 they used for the ammonia-only operation. The author also observed that for the stable operation of the engine, the minimum compression ratio was highly dependent on the cetane number of secondary fuels like diesel. The author also observed that satisfactory combustion could only be observed when ammonia was injected into the cylinder no later than 40°CA before the end of the diesel injection. Meanwhile, late ammonia injection, like slightly before the end of diesel injection, resulted in misfiring. Recent tests on a turbocharged engine showed that mixing ammonia and diesel affected fuel consumption and emissions. Low diesel content made engine less efficient, while high ammonia content increased fuel consumption. The best efficiency was found with 40–60% diesel energy. More ammonia lowered flame temperatures, which reduced NOx emissions but increased HC and CO emissions. Soot emissions were low, with more ammonia, and CO_2 emissions were also reduced. However, unburned ammonia was high, likely due to incomplete combustion. Biodiesel showed similar trends to regular diesel but with slightly more NOx and lower HC emissions. Another method tested involved using a small amount of diesel to ignite large amounts of ammonia. This approach achieved the same power as diesel alone but had poor fuel efficiency and emissions due to incomplete combustion.

Several injection strategies have been used to optimize the engine performance using ammonia as fuel. **Direct Injection (DI)** involves injecting ammonia directly into the combustion chamber at high pressure, offering high thermal efficiency and precise control over fuel quantity and timing, though it demands complex, high-cost systems capable of handling ammonia's corrosive nature. **Port Fuel Injection (PFI)** simplifies the fuel delivery system by injecting ammonia into the intake manifold, improving fuel–air mixing but sacrificing efficiency and increasing the risk of pre-ignition and backfire. **Dual-fuel injection** combines ammonia with a secondary fuel, such as diesel or hydrogen, to stabilize ignition, making it compatible with existing engines; however, it relies on fossil fuels or hydrogen, reducing ammonia's carbon-free advantage. **Pilot Ignition** uses a small quantity of diesel or another reactive fuel to ignite ammonia effectively, especially under low-load conditions, but its dependence on fossil fuels limits carbon reduction. Among these techniques, DI

Table 8.2 Injection strategies and their advantages and limitations

Techniques	Advantages	Limitations
Direct injection	High efficiency, reduced pre-ignition risk	High system cost, ammonia handling challenges
Port fuel injection	Simple system, better fuel–air mixing	Lower efficiency, higher knock risk
Dual-fuel injection	Stable ignition, compatible with existing engines	Requires secondary fuel, complex control
Pilot ignition	Stable ignition, good for multi-load conditions	Fossil fuel reliance, limited CO_2 reduction potential

is the most promising for achieving high thermal efficiency and precise combustion control despite its complexity and cost. For simpler applications, Dual-Fuel Injection with hydrogen as a secondary fuel balances combustion stability with reduced emissions, offering a practical alternative. Table 8.2 shows different injection strategies and their advantages and limitations in engine applications.

8.4 Engine Modifications Required for Ammonia Application

Ammonia dual-fuel engines need a separate fuel storage and supply system, posing a substantial change to the existing diesel engine setup. Ammonia is a gaseous fuel and has different physicochemical properties from its counterpart. Thus, the fuel line assembly components must be selected, keeping these things under consideration. In the case of a dual-fuel diesel-ammonia engine, ammonia fuel is injected similarly to gasoline in the intake port. However, various alternate injection approaches are possible, such as liquid ammonia injection in the port and direct injection of gaseous ammonia into the combustion chamber, which is not popular due to its complexity. Apart from the fuel assembly line, the engine compression ratio can be suitably modified to harness the octane number advantage of ammonia in the case of gasoline engines. In the case of a compression ignition engine, the auto-ignition property of ammonia must be considered while optimizing the engine's compression ratio. Figure 8.2 shows the ammonia fuel injection in a direct-injection diesel engine (Bjørgen et al. 2024).

Ammonia's high auto-ignition temperature (924 K) makes it difficult to ignite under standard diesel engine compression ratio range of 16:1 to 18:1. To address this; the compression ratio must be increased to levels as high as 27:1 (Schönborn 2020; Dinesh and Kumar 2022) or more, ensuring that the air–fuel mixture reaches the necessary temperature for ignition during compression. However, such adjustments need to be carefully balanced to avoid excessive mechanical stress and thermal

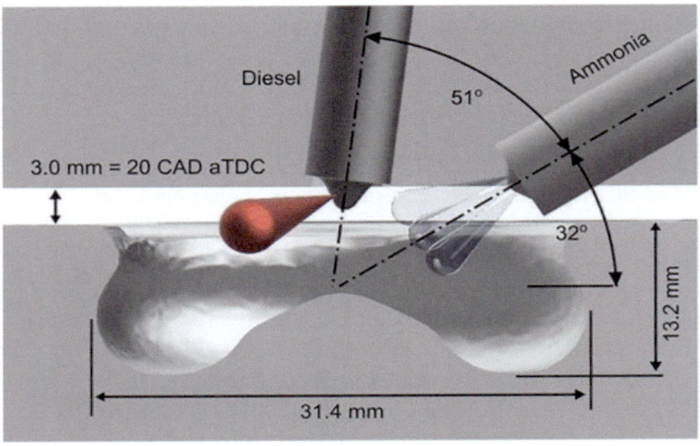

Fig. 8.2 Direct ammonia injection into the combustion chamber (Bjørgen et al. 2024)

loads on engine components, which could compromise engine durability and reliability. In addition to compression ratio adjustments, ignition assistance is essential for ammonia-fueled engines due to its low reactivity and narrow flammability range. Ammonia often requires an external ignition source to ensure stable combustion. Pilot ignition is a widely used approach, where a small amount of a highly reactive fuel, such as diesel or hydrogen, is injected to ignite the ammonia-air mixture (Liu et al. 2024). This method leverages the high reactivity of the pilot fuel to overcome ammonia's ignition challenges (Xu et al. 2023). Advanced technologies, such as plasma-assisted or laser ignition, are also being investigated for their ability to enhance ignition reliability in ammonia-fueled engines. These modifications to the compression ratio and ignition assistance systems are critical for enabling the efficient and reliable use of ammonia in diesel engines.

8.5 Performance, Combustion, and Emissions Characterizations

High Pressure Direct Injection (HPDI) is a high-performance fuel injection technique in which ammonia is directly injected into the combustion chamber under high pressure to ensure efficient and controlled combustion. HPDI systems employ specially designed solenoid-based direct injectors that maintain accurate flow rates and provide precise injection control for gaseous ammonia. These injectors are engineered to handle ammonia's low energy density and high ignition temperature while ensuring reliable operation under the extreme conditions of high-pressure injection. The injectors must also be constructed from materials resistant to ammonia's corrosive properties, ensuring durability and safety over extended operation.

In some advanced implementations, a single solenoid injector is designed with a co-axial configuration, allowing for the simultaneous or sequential injection of ammonia and diesel fuel. This innovative design integrates the delivery of ammonia and a pilot fuel, such as diesel, in a single compact unit. The co-axial injector design optimizes combustion stability by leveraging the high reactivity of diesel for ignition assistance while maintaining the primary role of ammonia as the main fuel. The seamless operation of such injectors requires highly advanced control algorithms in the engine's electronic control unit (ECU), ensuring precise synchronization of dual-fuel delivery to prevent knock, pre-ignition, and ammonia slip. The HPDI system's ability to finely control injection parameters allows it to adapt dynamically to varying load and operating conditions, offering superior performance compared to traditional injection methods. The integration of advanced injector designs, high-pressure pumps, and robust sealing mechanisms makes HPDI a technically sophisticated solution for enabling ammonia combustion in modern diesel engines. Continued innovation in injector technology and control systems is likely to further enhance the efficiency and feasibility of HPDI systems in ammonia-powered applications.

Several research have been done on the performance assessment of ammonia-diesel dual-fuel combustion in IC engines. The research focuses on the effect of ammonia energy fraction and diesel injection timing, which impacts the performance and emissions of ammonia-diesel dual-fuel (ADDF) engine. Increasing the ammonia energy fraction results in several key performance changes due to ammonia's unique properties. The peak cylinder pressure drops because ammonia has a high auto-ignition temperature, requiring a higher ignition temperature than fuels like diesel. This high ignition threshold delays the combustion process, causing lower peak pressure in the cylinder. The start of combustion (SOC) and overall combustion phasing are also delayed (Yousefi et al. 2022). Since ammonia ignites slowly, the engine must adjust its timing, to start the combustion process beginning of the expansion stroke. This shift impacts the timing and efficiency of the combustion. Lastly, the thermal efficiency of the engine decreases with higher ammonia fractions. Ammonia's low flame speed means the flame spreads through the air–fuel mixture slowly, leading to incomplete combustion and reduced energy extraction. Consequently, the engine experiences lower thermal efficiency as a significant portion of the ammonia-air mix does not burn completely by the end of the combustion cycle. These combined effects of reduced peak pressure, delayed combustion, and lower thermal efficiency are the challenges of using higher ammonia energy fractions on engine performance (Xu et al. 2023). Deng et al. (2024) studied an ammonia-diesel dual-fuel engine's combustion and emissions characteristics, varying high premixed ammonia fractions using numerical modeling. Increasing ammonia fractions (from 0.4 to 0.9) and keeping constant injection duration led to prolonged ignition delay, reduced in-cylinder pressure, and a drop in indicated thermal efficiency (ITE) by 25%. The major drop in ITE occurs while increasing the ammonia energy fraction from 0.8 to 0.9. In contrast, the ITE increased with increasing ammonia energy fraction from 0.4 to 0.9 through higher diesel injection pressure at constant injection rate settings. Bjorgen et al. (2024) examined the combustion dynamics of a compression ignition engine operating in high-pressure dual-fuel mode with liquid ammonia and diesel. The liquid ammonia was directly injected into the combustion chamber with a gasoline direct injector at 180 bar injection pressure for ammonia. The study was conducted by keeping fixed diesel injection timing (−15° CA bTDC) and varying the ammonia injection timing (−80°–2.5° CA aTDC) and ammonia energy share (40, 50, and 60%). For early ammonia injection, ammonia gets premixed with air and does not combust properly due to flame extinction in intricate regions, leading to unburnt ammonia emissions, also known as ammonia slip. Meanwhile, late injection increases ignition delay by accumulating large ammonia vapors. The optimum injection timing for ammonia was found close to diesel injection. Wang et al. (2023) investigated the impact of hydrogen blending in ammonia on diesel engine combustion and emission performance. A hydrogen blending ratio of 30% showed a 10.7% rise in peak in-cylinder pressure due to hydrogen's higher reactivity than pure ammonia. This improvement in combustion dynamics lead to 1.8% increase in power output, demonstrating the potential of hydrogen-ammonia blends to boost engine performance. However, the study showed an increase in NOx emissions by 58.8% with higher hydrogen content, a common challenge in combustion systems utilizing hydrogen. The study concludes that blending hydrogen with ammonia in diesel-ignited engines is a promising approach to enhance combustion efficiency and power output, but further research is needed to mitigate the associated increase in NOx emissions, ensuring that the environmental benefits of using zero-carbon fuels like ammonia are fully realized. Zhang et al. (2024) studied the effect of liquid ammonia high-pressure direct injection (HPDI) on ammonia-diesel dual-fuel heavy-duty engines. The liquid ammonia HPDI strategy improved fuel–air mixing, leading to more thorough combustion. The results showed that replacing 80% of the fuel energy with ammonia increased the indicated mean effective pressure (IMEP) by 8.9% and the ITE by 10.6%. The setup also modifies the timing and direction of ammonia injection, allowing for finer control over combustion and emission formation, significantly reducing unburned ammonia emissions even at high ammonia energy fractions. Liu et al. (2024) evaluated a CRDI ammonia/diesel dual-fuel engine for combustion performance, emissions, and fuel economy. The results showed that increasing the ammonia energy fraction (AEF) to 60% led to a significant deterioration in combustion efficiency, with in-cylinder pressure reduction and ignition delay increment. At 60% AEF, CO_2 emissions decreased by 50.2%, but cyclic fluctuations reached 5.06%. Increasing diesel injection pressure (DIP) from 100 to 130 MPa reduced NH_3 emissions by 20.8% and improved combustion efficiency, though further increase to 140 MPa led to incomplete combustion and reduced BTE. Advancing diesel injection timing (DIT) from − 4°CA ATDC to − 12°CA ATDC reduced CO, HC, and NH_3 emissions by 71.3%, 52.7%, and 57.6%, respectively, while increasing N_2O emissions by 19.9%. These findings highlight the trade-offs between emissions and efficiency in the ADDF engine and suggest that optimized injection strategies could enhance performance while mitigating emissions. Ammonia share was found optimum up to 84.2% of the input energy by Nadimi et al. (2023), which increases the ITE by 5.6% and reduces in exhaust gas temperature (EGT) by 132 °C. As the ammonia energy share (AES) increased, combustion shifted from diffusion to premixed mode, leading to reductions in combustion duration and phasing by 6.8° CA and 32° CA, respectively. At the highest AES, peak pressure rise rate (PRR) increased to 9.5 bar/deg, 76.5% higher than diesel-only operation. Replacing more than 35.9% of diesel with ammonia resulted in a decrease in greenhouse gas (GHG) emissions from 727 to 243 g/kWh, despite a significant increase in

N_2O emissions. Wang et al. (2024) explored the use of natural gas as a combustion enhancer for ammonia fuel engines through simulation research. Mixing natural gas with ammonia significantly improved combustion efficiency and power output, with the optimal mixture ratio found to be A50/N50 with 10% diesel energy. The engine's maximum in-cylinder pressure and power increase with the ammonia natural gas ratio, peaking at A80/N20. While natural gas enhances ammonia combustion and reduces ammonia liberation, it also raises CO_2 and NOx emissions. As natural gas proportion increases beyond 70%, CO and soot emissions rise significantly, though N_2O emissions decrease. The engine's effective power decreases when the natural gas ratio exceeds 70%, and the overall emission performance worsens despite the reduced N_2O levels. Natural gas, like hydrogen, promotes ammonia combustion but at the cost of increased NOx and CO_2 emissions. Bro et al. (Bro and Pedersen 1977) did a performance comparison for different alternative fuel blend applications in the diesel engine. Ammonia was not found suitable due to its low combustion, longer ignition delay, and ammonia emissions. Nadimi et al. (2022) studied the effect of ammonia-biogas mixture with biodiesel as the pilot fuel in a single-cylinder diesel engine. The brake thermal efficiency of ammonia-biodiesel was found to be lower compared to pure biodiesel combustion due to unburned ammonia. The brake thermal efficiency for biodiesel peaks at 32.27% at 88% load and declines at full load due to high exhaust gas temperatures. Ammonia reduces exhaust gas temperature by 100 K at full load. Brake-specific fuel consumption (BSFC) is higher with ammonia due to its lower heating value (LHV). Figure 8.3 illustrates the impact of ammonia-biogas mixture in biodiesel engine performance parameters.

As the ammonia energy share (AES) increases (Fig. 8.4) to 61.6%, the peak in-cylinder pressure rises, reaching 86.2 bar compared to 76.3 bar for pure diesel operation. However, the peak pressure declines when the AES increases to 84.2%. This indicates that the peak in-cylinder pressure initially increases with higher ammonia substitution but decreases at very high AES levels (Nadimi et al. 2023). In another study (Tutak et al. 2024), as ammonia shares increased up to 34%, maximum pressure increased compared to the reference pressure (D100), not exceeding 0.1 MPa was observed. The maximum pressure also occurred earlier, closer to TDC. Such an ammonia share led to an increase in the rate of pressure rise $dp/d\varphi$. It was noted that for the entire range of ammonia shares combusted with diesel fuel, the values of $(dp/d\varphi)_{max}$ were higher than for the reference fuel.

Nadimi et al. (2023) found a reduction in carbon dioxide (CO_2) emissions, dropping from 7.0% to 0.9% with an increase in ammonia energy share from 0 to 84%. This reduction is attributed to ammonia's lack of carbon content, eliminating CO_2 production. Concurrently, carbon monoxide (CO) emissions have been reduced markedly, from 7592 ppm in pure diesel mode to only 140 ppm, with the highest 84% ammonia energy share. However, ammonia combustion increased nitrogen oxide (NOx) emissions, rising from 831 ppm in diesel-only operation to 2359 ppm at 84% AES. This is attributed to ammonia's combustion characteristics, particularly its lower flame temperature and slower ignition, which affect NO formation in the engine. Additionally, the introduction of ammonia increased nitrous oxide (N_2O) emissions, a potent greenhouse gas with a high global warming potential (GWP). N_2O

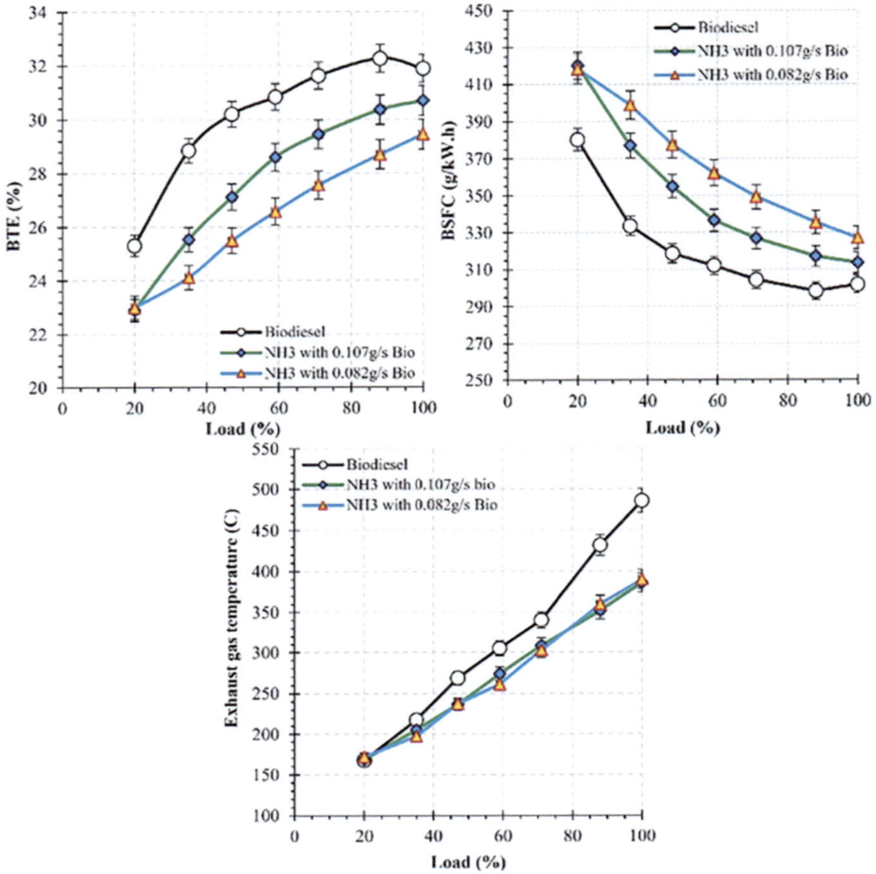

Fig. 8.3 Comparison of performance parameters for ammonia-biogas and pure biodiesel engine performance under different loads (Nadimi et al. 2022)

levels rose from negligible in the diesel-only mode to 90 ppm with ammonia before decreasing at 84% AES. Despite the lower CO_2 emissions, the high GWP of N_2O can offset the climate benefits. The study also found that ammonia significantly increased unburned ammonia emissions, from 7 ppm in diesel-only operation to 14,800 ppm at the maximum AES, due to incomplete combustion and ammonia's low flame speed. Furthermore, particulate matter (PM) emissions were notably reduced with higher ammonia levels, dropping from 22 mg/m^3 in diesel mode to just 4.7 mg/m^3 with the highest AES, indicating fewer soot particles due to ammonia's carbon-free nature. While ammonia substitution decreases CO_2 and PM emissions, it presents challenges with increased NO and N_2O emissions, necessitating further optimization and technological improvements to realize its environmental benefits fully. The effect of ammonia energy share on CO_2, H_2O, NO_2, and N_2O emissions is presented in Fig. 8.5 (Nadimi et al. 2023).

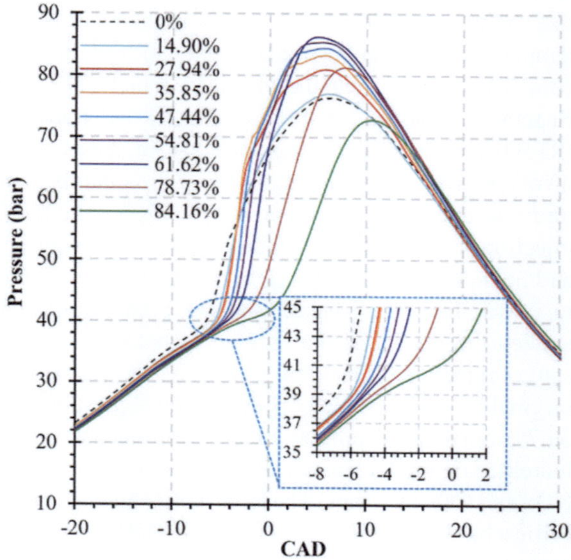

Fig. 8.4 In-cylinder pressure variation for different ammonia energy under dual-fuel mode (Nadimi et al. 2023)

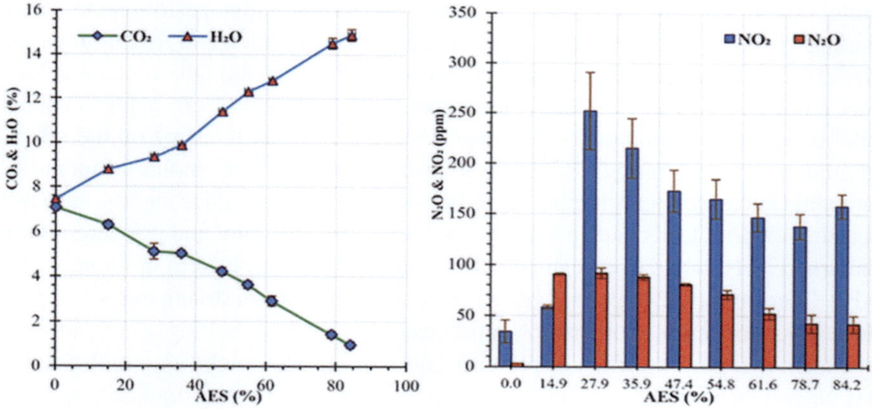

Fig. 8.5 CO_2, H_2O, NO_2, and N_2O emissions from ammonia-fueled diesel engines at various ammonia energy fractions (Nadimi et al. 2023)

As more diesel fuel is replaced with ammonia, carbon monoxide (CO) emissions decrease significantly as shown in Fig. 8.6, dropping from 7592 ppm in pure diesel operation to just 140 ppm. This reduction is because ammonia has no carbon, so it doesn't produce CO. However, ammonia can increase nitrogen oxide (NO) emissions. Initially, NO levels drop when a small amount of ammonia is used, falling from 831 ppm in diesel-only operation to 491 ppm with 14.9% ammonia. This decrease is due to ammonia's lower combustion temperature, which reduces thermal NO formation. However, as the ammonia proportion increases further, NO emissions rise sharply to 2359 ppm. This increase happens because ammonia's higher ignition temperature and combustion characteristics promote NO formation, especially when combustion starts near the top dead center (TDC). Wu et al. (2024) investigated Reactivity Assisted Jet Ignition (RAJI) and Reactivity Turbulent Jet Disturbance (RTJD) approaches for ammonia use in diesel engines. The RAJI mode, which involves early diesel injection into a pre-chamber, improved in-cylinder temperature and reduced N_2O emissions by 13.56%. The RTJD mode, involving later diesel injection, reduced N_2O, unburned NH3, and soot emissions by 28.81%, 27.79%, and 35.08%, respectively, while improving gross indicated thermal efficiency by up to 1.08%. In a CRDI-based dual-fuel ammonia-diesel engine, Nadimi et al. (2024) studied the impact of the number of nozzles in the biodiesel injector and various start of injection (SOI) timings. Results revealed that welding three nozzles from the original six-nozzle injector achieved the highest indicated thermal efficiency (ITE) of 39.7% and reduced ammonia and CO emissions by 29.2%. Late injection of both fuels increased particulate matter (PM) emissions from 10.5 to 15.2 mg m^{-3}, but decreased NOx and CO emissions. The lowest N_2O emission of 115.0 ppm was observed at the earliest biodiesel SOI of − 24 CAD. Shifting the biodiesel SOI from − 24 to − 14°CA improved combustion efficiency, reducing ignition delay and combustion duration. The study also found that reducing the number of injector nozzles lowered NOx emissions from 4.1 to 2.0 g/kWh and improved overall engine performance. These findings indicate that optimizing injector nozzle configuration and SOI timings can significantly enhance the performance and emissions of ammonia-biodiesel dual-fuel engines. Chen et al. (2024) investigated the emissions from a four-cylinder pilot diesel-ignited ammonia-diesel dual-fuel (DF) engine, focusing on the effects of ammonia energy ratio (AER), engine load, and speed. The amount of unburned ammonia increases with higher AER but decreases with higher engine load and speed. At higher loads and speeds, the in-cylinder temperature and pressure reduce unburned NH_3, while over-lean conditions at higher speeds lead to more unburned NH_3. Chen et al. (2024) also suggested some emission control strategies such as pilot diesel injection timing, combustion optimization, liquid flash-boiling spray, and after-treatment system. The effect of pilot injection timing and combustion optimization has already been discussed. Implementing a liquid flash-boiling ammonia spray at the intake manifold can enhance charging efficiency and improve combustion characteristics. Various after-treatment technologies such as selective catalytic reduction (SCR), diesel oxidation catalysts (DOC), and ammonia slip catalysts (ASC) can be used to reduce emissions effectively.

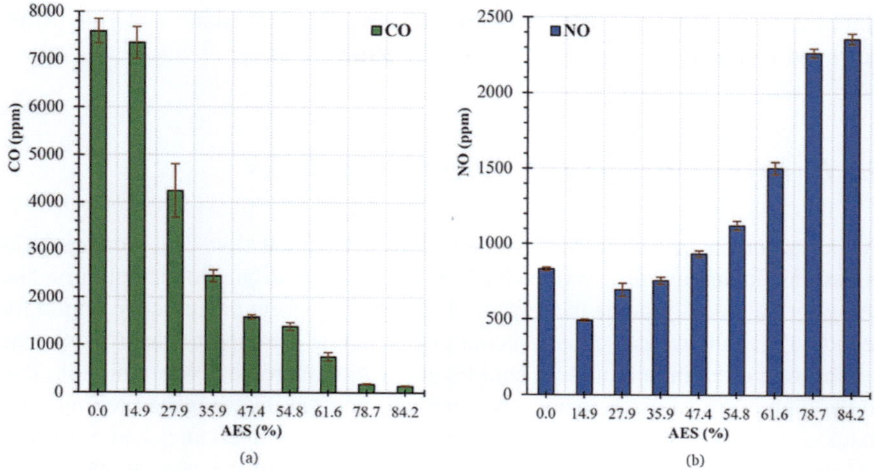

Fig. 8.6 CO and NO emissions from ammonia-fueled diesel engines at various ammonia energy fractions (Nadimi et al. 2023)

8.6 Challenges Associated and Its Solution Approach

Ammonia diesel dual-fuel engines face many challenges in obtaining optimum combustion performance. Low flame speed and high auto-ignition temperature of the ammonia are critical to utilize fully in the existing engine. Thus, engine's combustion efficiency depends on several critical parameters like injection timing, ammonia-diesel ratio, heat release rate, etc. A higher heat release rate (HRR) indicates complete combustion, which improves efficiency (Qian et al. 2024). On the other hand, if the HRR is too high, higher combustion temperatures may cause more NOx emissions. The timing of diesel fuel injection is also crucial, as early injection can enhance fuel–air mixing but may lower peak pressures and heat release rates, while late injection can delay ignition and lead to incomplete combustion. Additionally, the proportion of ammonia in the fuel mix significantly affects efficiency, with an ammonia energy ratio between 40 and 60%, often yielding optimal performance. NH_3 proportion in fuel mixture significantly affects the efficiency, which is 50 to 65%, showing optimum performance. A higher percentage of ammonia may increase NOx emissions due to higher combustion temperature. Ammonia's high ignition energy and low flame speed can also reduce combustion efficiency; however, these qualities can be improved by using pilot diesel injections, enhancing swirl intensity, and improving combustion chamber design. Optimizing these factors is necessary to maximize combustion efficiency in dual-fuel ammonia-diesel engines. Ammonia blending with PODE (polyoxymethylene dimethyl ether) can enhance ignition quality. The results from the research showed that blending ammonia with fuels such as PODE can enhance ignition characteristics (Mao et al. 2024). PODE exhibits an earlier ignition, a wider flame area, and a shorter combustion duration due to its higher cetane number and greater injection mass at the same energy input. Modification of the combustion chamber geometry can also enhance the air motion inside the piston bowl and will help to enhance the fuel–air mixing process.

8.6.1 NOx Emissions and Mitigation Strategies

While ammonia is carbon-free, studies show its combustion can produce NOx emissions, including nitric oxide (NO), which may contribute to air pollution and smog formation. Ammonia diesel dual-fuel engines present an attractive option for reducing greenhouse gas emissions but also generate nitrogen oxides (NOx). The combustion of ammonia in these engines can lead to increased NOx emissions due to the high combustion temperatures. Another important factor is the fuel injection method. Advanced injection time can raise combustion temperatures and pressures, which raises NOx emissions; on the other hand, optimized injection strategies can enhance air–fuel mixing and perhaps lower emissions. The unburned ammonia can increase nitrous oxide (N_2O) emissions, a greenhouse gas. Changing fuel injection methods, such as split and pilot fuel, can improve ammonia-air mixing and enhance combustion, lowering NOx production. Furthermore, lowering combustion temperatures may help to minimize thermal NOx emissions by modifying engine design and improving combustion chamber shape and thermal management.

8.6.2 Safety and Handling of Ammonia

Although ammonia has many uses, it's poisonous and combustible properties represent a severe risk to the environment and all living things. The safe inhalation limit of NH_3 varies from human to human. The report (Nielsen 1995) shows that the lower limit of NH_3 inhalation is from 0.4–2 ppm wt, over which biopotential damage to the brain can occur. An exposure of 9 to 45 ppm can cause eye and nose irritation. According to OSHA (Occupational Safety and Health Administration), the permissible exposure level (PEL) has been set at 50 ppm over 8 h of exposure (U.S. Government Printing Office n.d.). Thus, safety and handling of ammonia is crucial. Ammonia has been handled safely for many decades, but we also have the technology to lower the safety risk further through continuous improvement in handling and storage. Due to the corrosive property of NH_3, it can only be stored in a steel or plastic tank. Strict protocols should be followed to handle ammonia. The use of personal protective equipment (PPE), safe handling techniques, and awareness of emergency response procedures are necessary for ensuring the safety of workers and the surrounding environment. Liquid ammonia is transported in tankers, ships, rail, cars, and other road transportation. The report (Hignett 1985) states that for ammonia to be transported through pipelines, it must be heated to a minimum of 2 °C. The pipelines frequently require heating the ammonia at the supply terminal and cooling it to − 33 °C at the receiving terminal. The PVT characteristics of ammonia with a low water content are crucial for ammonia transportation, especially by pipeline (Tvinnereim and Mehling 2018). The report shows that (Tan et al. 2020), between 2007 and 2017, about 60 ammonia accidents in different industries across the United Kingdom were reported. In India, Gangopadhyay and Das (2008) analyzed and discussed two ammonia leakage accidents in West Bengal in cold storage and ice-manufacturing plants due to the hose pipe and a valve failure. The high number of severe ammonia spill accidents has made chemical hazards a severe concern. Factors like equipment failures, operational issues, human errors, poor process design, and inadequate safety measures contribute most to these accidents. So, addressing these issues through better equipment, training, and safety protocols is necessary to prevent future accidents and protect workers and the environment.

Handling ammonia-based engines requires stringent safety measures due to ammonia's toxicity, corrosive nature, and flammability. Secure storage is essential, utilizing corrosion-resistant, pressure-rated tanks equipped with pressure-relief valves and secondary containment systems to prevent leaks. Advanced ammonia sensors should be installed in storage and engine areas for early leak detection and mitigation. Engine components such as fuel lines, injectors, and seals must be constructed from corrosion-resistant materials and regularly maintained to ensure durability. Proper ventilation systems are critical to prevent ammonia accumulation and safely vent emissions. Personnel must be equipped with appropriate protective gear and receive comprehensive training on ammonia handling, including emergency response protocols. Additionally, emergency systems, such as shut-off valves, neutralizing agents, and detailed spill response plans, should be in place to address potential leaks promptly. These measures are vital for ensuring ammonia-powered engines' safe and efficient operation.

8.6.3 Hazard and Safety Aspects of Ammonia Engine

Ammonia is highly toxic and can pose serious health risks upon exposure. Inhalation of ammonia gas, even at low concentrations, can irritate the respiratory system, while higher concentrations can cause severe damage to the lungs and respiratory tract. Direct contact with liquid ammonia can result in chemical burns to the skin and eyes. Prolonged or high-level exposure may lead to systemic poisoning, which can be fatal. Proper safety measures, including ventilation, protective equipment, and leak detection systems, are essential to mitigate these risks. Operating ammonia-fueled engines require careful material selection to address ammonia's corrosive and reactive nature. Fuel system components, including tanks, lines, and injectors, must be made from ammonia-resistant materials such as stainless steel or nickel-based alloys. Non-metallic components like seals and gaskets should use corrosion resistant materials like PTFE or Viton to prevent degradation. High-pressure injection systems require precision components with hardened or coated surfaces for durability. Ammonia can also affect standard lubricants, necessitating the use of specialized ammonia-resistant oils. Exhaust systems and catalytic converters must be constructed from corrosion-resistant materials to handle ammonia slip effectively. Proper material compatibility is essential to ensure safety, reliability, and longevity of ammonia-powered engines.

8.7 Economic and Environmental Impacts

8.7.1 Cost Analysis

Ammonia cost analysis includes production, storage, transportation, and proper utilization. The production of ammonia is costly for several reasons, including energy, feedstock (primarily natural gas for hydrogen production), cost of capital (plant development and maintenance), people and catalyst costs, transportation, and compliance with environmental regulations. Costs associated with feedstock, especially natural gas, can make up 50–70% of overall production expenses. Ten to twenty percent of the overall cost comprises energy, capital, labor, maintenance, and regulation costs. However, initially more expensive, renewable energy sources provide sustainable alternatives for manufacturing ammonia. So, technological developments and process improvement can lower overall costs and increase efficiency.

Shiozawa (2020) assessed the economics of manufacturing CO_2-free ammonia using chemical processes as part of a study funded by the Institute of Energy Economics in Japan. He found that ammonia produced by using a combination of natural gas and carbon capture and sequestration (CCS) will cost roughly \$350 per ton by 2030. The ton of CO_2-free ammonia using renewable hydrogen will cost roughly \$750. Bartels (2008) conducted an economic assessment of the shipping and storage costs of hydrogen fuel and ammonia. The study found that ammonia was significantly cheaper than hydrogen (0.54 compared to 14.95 \$/kg of H_2) for 182 days of storage. Also, the pipeline transportation costs for ammonia were less than those for hydrogen (0.194 compared to 0.51 − 3.22 \$/kg of H_2) based on a transport distance of 1610 km. Nadimi et al. (2023) have done a cost analysis of an ammonia-diesel dual-fuel engine, analyzing the financial impact of indicated power (P_i) by comparing the ammonia dual-fuel mode with pure diesel operation. The engine uses 264 g/kWh of fuel while operating on pure diesel, but it uses 432.6 g/kWh of ammonia and 35.6 g/kWh of diesel when operating on dual fuel. The P_i cost is much lower than that of the pure diesel mode, even though the ammonia usage is more extensive. The cost of P_i in the ammonia/diesel engine is about 0.24, which is 44% less than the conventional diesel engine, as shown in Table 8.3.

Table 8.3 Cost analysis of ammonia-diesel dual-fuel mode with only diesel operation (Nadimi et al. 2023)

Operating mode	Fuel	Fuel price (€/kg)	ISFC (g/kWh)	Indicated power cost (Pi)
Diesel only	Diesel	1.65	264	0.43
Ammonia-diesel dual	Ammonia (84.16%)	0.43	432.67	0.24
	Diesel (15.84%)	1.65	35.64	

8.7.2 *Environmental Benefits and Drawbacks*

Ammonia is a suitable alternative to conventional fossil-based fuels due to its several environmental advantages. The primary advantage of ammonia lies in its zero-carbon content, which eliminates the carbon emissions in diesel engines, unlike conventional fuels, which emit carbon dioxide and carbon monoxide. However, the combustion of ammonia generates only nitrogen and water vapors upon complete combustion. Additionally, ammonia can be generated using renewable energy sources, known as green ammonia. In this process, renewable energy sources such as solar, wind, and hydropower can generate electricity to split water into hydrogen, unlike steam thermal reformers. Further, the generated hydrogen can form ammonia using the Haber–Bosch process. The high energy density of ammonia makes it a suitable choice for long-distance transportation and an effective energy carrier for hydrogen. The existing storage and infrastructure require minimal economic requirements to switch to ammonia fuel. The liquid phase ammonia can also be utilized suitably in diesel engines with minimum modifications. Moreover, ammonia combustion engines do not release particulates and sulfur oxide emissions, reducing health hazards.

Although ammonia possesses several advantages in terms of combustion and emissions in diesel engines, it still need a solution in specific emergencies. The major drawback of ammonia is the occurrence of health hazards in direct contact with skin or inhalation of ammonia. Several precautions regarding handling, storage, and transportation should be kept under check, maintaining stringent security standards. Renewable energy sources are very scarce at the current time and need a lot of investment and research to make them viable for hydrogen production and their utilization in ammonia production to make them economically feasible. Substantial financial support from various sectors and organizations, as well as technological progress, is desired. The high energy density of ammonia supports transportation compared to hydrogen. However, the desired power needs a high amount of ammonia usage to get power similar to that of a diesel engine, which will impact the diesel engine's efficiency. Moreover, the combustion of ammonia can produce nitrogen oxides (NOx), dangerous pollutants that damage respiratory conditions and air quality. Using catalytic converters and advanced combustion technology to reduce NOx emissions increases the expense and complexity of using ammonia as a fuel.

Lastly, compared to the infrastructure for fossil fuels, the current setup for ammonia distribution and refilling is lacking. Ammonia needs to be developed with significant infrastructure investment to become a widely available and practical fuel.

8.8 Future Trends and Research Opportunities

A strong infrastructure for ammonia manufacturing must be established, combined with renewable energy sources such as hydropower, solar electricity, or wind, that will reduce the overall carbon footprint. Ammonia application in diesel engines requires a robust fuel supply and refueling infrastructure, transportation, and distribution grids. The major obstacle in this regard is the storage container and its capacity, as ammonia is corrosive and needs special design care in storage tanks based on liquid or gaseous storage form. There is a significant research and development requirement in storage infrastructure, transport policy, and optimization of economic feasibility. The addition of ammonia to diesel engines in dual-fuel mode offers challenges such as longer combustion duration, low fuel efficiency, and higher CO & HC emissions due to the low combustion temperature of the diesel fuel mixture. In addition, the NOx emissions were higher in dual-fuel mode due to the prolonged ignition delay period. The ignition delay in the dual-fuel operation could be solved using highly reactive fuels such as hydrogen or nano-particles and ammonia. The research on adding oxygenated additives, aluminum oxide, metal oxide, and Ruthenium oxide, along with biodiesel (spirulina micro-algae, Aleurites moluccanus, Castor oil) in replacing the diesel engine with fossil-free engines is a hot topic of interest. These blends have the potential to solve energy density, storage, and combustion stability problems, making them a promising area for future research and development in the quest for sustainable, clean energy solutions. The NOx emissions could be reduced by using strategies such as lean burn, exhaust gas recirculation (EGR), and selective catalytic reduction (SCR). The catalytic converter for ammonia combustion should be explored for a suitable metal oxide coating for NOx reduction, effective at a large range of temperatures. Research is also underway to explore advanced combustion technologies such as reactivity-controlled compression ignition (RCCI) and homogeneous charge compression ignition (HCCI) for ammonia applications for higher efficiency. Further research is needed to improve catalysts for ammonia cracking, which is crucial for onboard hydrogen production. Efficient and resilient catalysts are necessary to enhance reaction rates and optimize the selective generation of products during ammonia decomposition.

Green ammonia is a clean and potential ammonia source, where ammonia is produced utilizing renewable energy to reduce the carbon footprint. It is a clean, sustainable energy source with much potential. It provides a more environmentally friendly method of making fertilizers for farming. By attracting investment and generating new jobs, the development of green ammonia technology can strengthen the economy. Green ammonia is expected to be a major player in the transition to cleaner energy sources thanks to increased demand and government backing.

8.9 Policy Framework

The Japanese government aims for net-zero carbon emissions by 2050, incorporating ammonia as a critical component (Oshiro et al. 2018). The ITOCHU's "Integrated Project" of ammonia-fueled ships is one such initiative that explores ammonia as a marine fuel (Taylor 2020). Japan has set targets for ammonia's role in its energy supply, aiming for a 20% co-firing ratio of ammonia with fossil fuels by 2025, increasing to 50% by 2030, and ultimately achieving 100% ammonia-fueled internal combustion engines by 2050, with the option for carbon capture and storage (CCS) (Bloomberg 2018). To support these goals, Japan offers incentives like subsidies and tax breaks, fostering public–private partnerships for research, engine technology development, and infrastructure. The country is also creating a regulatory framework to ensure the safety, efficiency, and interoperability of ammonia-fueled engines, focusing on standardization and demonstration projects to test their feasibility in diverse applications (Brown 2021). Similarly, South Korea is heavily invested in ammonia's potential, particularly for use in the maritime sector. The country's 2050 Carbon Neutrality Vision includes ammonia as a key alternative fuel, with significant research being directed toward ammonia-powered ships and power generation. In the United States, the Department of Energy (DOE) is funding research into ammonia's production, combustion characteristics, engine designs, and emission control technologies, with a focus on its potential in the aviation and transportation sectors (Atchison 2024). Additionally, the U.S. Navy is looking into ammonia as a fuel for naval ships and submarines to support energy security and sustainability goals (Ammonia-Powered Engine to be Developed for Medium-Speed Marine Application 2024). The Australian Renewable Energy Agency (ARENA) has funded several projects that explore ammonia as a viable fuel option (Ammonia-Production-From-Renewables 2024). With its large industrial base, China invests in ammonia as a hydrogen carrier and fuel for power generation, exploring ways to reduce emissions from its industries and transportation systems. Research institutions in China are examining ammonia's combustion properties and its potential to reduce carbon emissions across various sectors.

8.10 Conclusion

The exploration of ammonia as a viable alternative fuel for diesel engines reveal significant potential in pursuing sustainable energy solutions. The review demonstrates that ammonia's carbon-free composition offers a substantial environmental advantage by reducing greenhouse gas emissions, mainly CO_2, significantly contributing to global warming. Technological advancements, such as dual-fuel systems and direct ammonia injection, have shown promise in overcoming ammonia's inherent challenges, including its low energy density, narrow flammability limits, and high auto-ignition temperature. The performance and emissions analysis of ammonia-diesel dual-fuel engines indicates that while ammonia can effectively reduce CO_2 and soot emissions, it also poses challenges related to combustion efficiency and NOx emissions. The need for engine modifications, such as optimized injection timing and pressure, is critical to enhance combustion efficiency and minimize unburned ammonia and NOx emissions. Moreover, ammonia's existing production and storage infrastructure, coupled with ongoing innovations in ammonia synthesis and transportation, further solidify its feasibility as a sustainable fuel option. However, economic and safety considerations must be meticulously addressed to ensure the practical deployment of ammonia in the transportation sector.

In conclusion, while ammonia presents a promising pathway toward cleaner diesel engine technology, continued research and development are essential to optimize its use and address the challenges identified in this review. Future studies should focus on refining combustion techniques, improving engine modifications, and exploring the integration of ammonia with other alternative fuels to maximize its potential.

References

Ammonia-production-from-renewables 2024. https://arena.gov.au/projects/ammonia-production-from-renewables/

Atchison J (n.d) Wabash valley resources to receive $1.6 billion in DoE support n.d. https://ammoniaenergy.org/articles/wabash-valley-resources-to-receive-1-6-billion-in-doe-support/. Accessed 24 Nov 2024

Bartels JR (2008) A feasibility study of implementing an Ammonia Economy. Iowa State University

Bjørgen KOP, Emberson DR, Løvås T. Combustion of liquid ammonia and diesel in a compression ignition engine operated in high-pressure dual fuel mode. Fuel 360. https://doi.org/10.1016/j.fuel.2023.130269

Bloomberg. Japan's Ammonia-Coal Co-Firing Strategy a Costly Approach to Decarbonization, Renewables Present More Economic Alternative n.d. https://about.bnef.com/blog/japans-ammonia-coal-co-firing-strategy-a-costly-approach-to-decarbonization-renewables-present-more-economic-alternative/.

Bro K, Pedersen PS (1977) Alternative diesel engine fuels: an experimental investigation of methanol, ethanol, methane and ammonia in a DI diesel engine with pilot injection. SAE technical paper

Brown T (2021) Japan's Road Map for Fuel Ammonia 2021. https://ammoniaenergy.org/articles/japans-road-map-for-fuel-ammonia/

Change IC et al (2014) Contribution of working group III to the fifth assessment report of the intergovernmental panel on climate change. Mitig Clim Chang 1454:147

Chen R, Li T, Wang X, Huang S, Zhou X, Li S et al (2024) Engine-out emissions from an ammonia/diesel dual-fuel engine—the characteristics of nitro-compounds and GHG emissions. Fuel 362. https://doi.org/10.1016/j.fuel.2023.130740

Deng X, Zhao W, Ye L, Gong S, Zhao F, Li J (2024) Combustion and emissions characteristics of ammonia/diesel dual-fuel engine with high premixed fractions in different injection modes. Int J Hydrogen Energy 49:1232–1244. https://doi.org/10.1016/j.ijhydene.2023.10.199

Dimitriou P, Javaid R (2020) A review of ammonia as a compression ignition engine fuel. Int J Hydrogen Energy 45:7098–7118. https://doi.org/10.1016/j.ijhydene.2019.12.209

Dinesh MH, Kumar GN (2022) Effects of compression and mixing ratio on NH3/H2 fueled Si engine performance, combustion stability, and emission. Energy Convers Manag X 15:100269. https://doi.org/10.1016/j.ecmx.2022.100269

Dong D, Wei F, Long W, Dong P, Tian H, Tian J, et al. Optical investigation of ammonia rich combustion based on methanol jet ignition by means of an ignition chamber. Fuel 345:128202. https://doi.org/10.1016/j.fuel.2023.128202

Ammonia-Powered Engine to be Developed for Medium-Speed Marine Application 2024. https://www.man-es.com/company/press-releases/press-details/2024/10/22/ammonia-powered-engine-to-be-developed-for-medium-speed-marine-applications. Accessed 24 Nov 2024

Gangopadhyay RK, Das SK (2008) Ammonia leakage from refrigeration plant and the management practice. Process Saf Prog 27:15–20. https://doi.org/10.1002/prs.10208

Haber F (2002) The synthesis of ammonia from its elements Nobel Lecture, June 2, 1920. Resonance 7:86–94

Hignett TP (1985) Transportation and storage of ammonia BT—fertilizer manual. In: Hignett TP (ed) Springer Netherlands, Dordrecht, pp 73–82. https://doi.org/10.1007/978-94-017-1538-6_7

Ishaq H, Crawford C (2024) Review and evaluation of sustainable ammonia production, storage and utilization. Energy Convers Manag 300:117869. https://doi.org/10.1016/j.enconman.2023.117869

Koch E (1945) Ammonia–a fuel for motor buses. J Inst Pet 31:498

Kurien C Mittal M (2022) Review on the production and utilization of green ammonia as an alternate fuel in dual-fuel compression ignition engines. Energy Convers Manag 251. https://doi.org/10.1016/j.enconman.2021.114990

Li J, Lai S, Chen D, Wu R, Kobayashi N, Deng L et al (2021) A review on combustion characteristics of ammonia as a carbon-free fuel. Front Energy Res 9:760356

Liu J, Wang X, Zhao W, Sun P, Ji Q (2024) Effects of ammonia energy fraction and diesel injection parameters on combustion stability and GHG emissions characteristics in a low-loaded ammonia/diesel dual-fuel engine. Fuel 360. https://doi.org/10.1016/j.fuel.2023.130544

Mao J, Zhang Y, Ma Y, Ma X, Wang Z, Wang SS (2024) Optical diagnostic study on ammonia-diesel and ammonia-PODE dual fuel engines. https://doi.org/10.4271/2024-01-2362

McKIE D (1933) Joseph Priestley (1733–1804), chemist. Sci Prog 28:17–35

Nadimi E, Przybyła G, Emberson D, Løvås T, Ziółkowski Ł, Adamczyk W. Effects of using ammonia as a primary fuel on engine performance and emissions in an ammonia/biodiesel dual-fuel CI engine. Int J Energy Res 46:15347–15361. https://doi.org/10.1002/er.8235

Nadimi E, Przybyła G, Lewandowski MT, Adamczyk W. Effects of ammonia on combustion, emissions, and performance of the ammonia/diesel dual-fuel compression ignition engine. J Energy Inst 107. https://doi.org/10.1016/j.joei.2022.101158

Nadimi E, Przybyła G, Løvås T, Adamczyk W. Effects of biodiesel injector configuration and its injection timing on performance, combustion and emissions characteristics of liquid ammonia dual direct injection engine. J Energy Inst 114. https://doi.org/10.1016/j.joei.2024.101605

Nielsen A (1995) Ammonia storage and transportation-safety BT—ammonia: catalysis and manufacture. In: Nielsen A (ed) Springer Berlin Heidelberg, Berlin, Heidelberg, pp 329–46. https://doi.org/10.1007/978-3-642-79197-0_7

Nur Syamimi M (2018) Quantitative risk assessment of anhydrous ammonia leakage and on site emergency plan at Bintulu industrial park/Nur Syamimi Mustapa

Oshiro K, Masui T, Kainuma M (2018) Transformation of Japan's energy system to attain net-zero emission by 2050. Carbon Manag 9:493–501

Pattabathula V, Richardson J (2016) Introduction to ammonia production. Chem Eng Prog 112:69–75

Press Information Bureau G of IM of P & NG. Press Information Bureau 2022. https://pib.gov.in/Pressreleaseshare.aspx?PRID=1848345. Accessed 15 Aug 2024

Qian Y, Pei X, Zheng L, Mi S, Ju D, Zhou D et al (2024) Improving thermal efficiency of an ammonia-diesel dual-fuel compression ignition engine with the addition of premixed low-proportion hydrogen. Int J Hydrogen Energy 58:707–716. https://doi.org/10.1016/j.ijhydene.2024.01.235

Schönborn A (2020) Aqueous solution of ammonia as marine fuel. Proc Inst Mech Eng Part M J Eng Marit Environ 235:142–151. https://doi.org/10.1177/1475090220937153

Shiozawa B (2024) The cost of CO2-free ammonia. Ammon Energy Assoc. https://ammoniaenergy.org/articles/the-cost-of-co2-free-ammonia/. Accessed 15 Aug 2024

Tan W, Lv D, Guo X, Du H, Liu L, Wang Y (2020) Accident consequence calculation of ammonia dispersion in factory area. J Loss Prev Process Ind 67:104271. https://doi.org/10.1016/j.jlp.2020.104271

Taylor I. GLOBAL: Industry stakeholders team up to develop ammonia-fuelled ships 2020. https://www.bunkerspot.com/global/50429-global-industry-stakeholders-team-up-to-develop-ammonia-fuelled-ships.

Travis AS, Travis AS (2018) Non-BASF Ammonia Technologies. Nitrogen Capture Growth Int Ind 225–264

Tutak W, Jamrozik A, Grab-Rogaliński K, Pyrc M. Effects of ammonia energy fraction on combustion stability and emissions characteristics of naturally aspired industrial dual-fuel diesel engine. Energy Convers Manag 314:118698. https://doi.org/10.1016/j.enconman.2024.118698

Tvinnereim E, Mehling M (2018) Carbon pricing and deep decarbonisation. Energy Policy 121:185–189. https://doi.org/10.1016/j.enpol.2018.06.020

U.S. Government Printing Office (n.d.) Washington D 20402. Code of Federal Regulations Title 29 Part 1910, Section 1000 (Labor)

Wang B, Yang C, Wang H, Hu D, Wang Y (2023) Effect of diesel-ignited ammonia/hydrogen mixture fuel combustion on engine combustion and emission performance. Fuel 331:125865. https://doi.org/10.1016/j.fuel.2022.125865

Wang B, Yang C, Wang H, Hu D, Duan B, Wang Y (2024) Effects of combustion and emission performance of ammonia/natural gas engines ignited by diesel. Fuel 358. https://doi.org/10.1016/j.fuel.2023.130323

Wu B, Shi M, Jin S, Wang J, Zi Z, Yang P et al (2024) Comparison and optimization of strategies for ammonia-diesel dual-fuel engine based on reactivity assisted jet ignition and reactivity turbulent jet disturbance under high load conditions. Int J Hydrogen Energy 67:487–499. https://doi.org/10.1016/j.ijhydene.2024.04.172

Xu L, Xu S, Bai X-S, Repo JA, Hautala S, Hyvönen J (2023) Performance and emission characteristics of an ammonia/diesel dual-fuel marine engine. Renew Sustain Energy Rev 185:113631. https://doi.org/10.1016/j.rser.2023.113631

Yang M, Hunger R, Berrettoni S, Sprecher B, Wang B (2023) A review of hydrogen storage and transport technologies. Clean Energy 7:190–216

Yousefi A, Guo H, Dev S, Liko B, Lafrance S (2022) Effects of ammonia energy fraction and diesel injection timing on combustion and emissions of an ammonia/diesel dual-fuel engine. Fuel 314:122723. https://doi.org/10.1016/j.fuel.2021.122723

Zhang Z, Di L, Shi L, Yang X, Cheng T, Shi C. Effect of liquid ammonia HPDI strategies on combustion characteristics and emission formation of ammonia-diesel dual-fuel heavy-duty engines. Fuel 367. https://doi.org/10.1016/j.fuel.2024.131450

Chapter 9
A Comprehensive Review on the Emission Characteristics of Ammonia Powered Engines

Md Shadab Reza, Jai Gopal Gupta, and S. Anil Lal

Abstract The development and investigation of various alternative fuels are essential due to the uncertain future of fossil fuels and the significant exhaust emissions from their combustion in internal combustion engines. The issue of global warming, due to the emission of greenhouse gases from fossil fuel combustion, is also a serious concern for the planet. Ammonia and hydrogen are examples of carbon-zero alternative fuels that reduce the amount of gaseous and particulate emissions from internal combustion engines. These fuels have been tested in their pure forms as well as blended with fossil fuels. Ammonia (NH_3) is an attractive carbon-free alternative fuel that reduces CO_2, SO_X, soot, and hydrocarbon pollutants. Its high-octane number permits the advancing of spark timing which offers an excellent knock resistance property. Also, the ease of production, storage, and transportation makes it a viable option. The complete combustion of ammonia produces only nitrogen and water, but due to oxidation of N_2 at elevated temperatures, it produces thermal NO_X. The challenges such as high N_2O, unburned ammonia, and hydrogen cyanide (HCN) emissions due to fuel-bound nitrogen must be addressed for effective ammonia combustion in internal combustion engines. Studies indicate that a blending ratio of 40% ammonia and 60% diesel in a dual-fuel method lowers NO_X emissions compared to pure diesel combustion, also advancing fuel injection timing and multiple fuel injections reduces the N_2O and NH_3 emissions. The blending of 50% ammonia and 50% CNG reduces the CO_2 emissions by 28%. Hydrogen cyanide emissions can be mitigated by advancing the combustion phase closer to TDC and heating the intake air. Further, with the application of after-treatment devices the emission of NO_X and unburned NH_3 can be minimized. This chapter provides a detailed review of the emission characteristics of ammonia-powered engines.

Keywords Alternative fuel · Green ammonia · Emission characteristics · Unregulated emissions

---

M. S. Reza · J. G. Gupta (✉) · S. Anil Lal
Department of Mechanical Engineering, National Institute of Technology Sikkim, Ravangla, Sikkim, India
e-mail: jaig@nitsikkim.ac.in

© The Author(s), under exclusive license to Springer Nature Singapore Pte Ltd. 2026
H. Valera and A. K. Agarwal (eds.), *Ammonia and Low Carbon Combustion Engine Technologies for De-Fossilizing Transport Sector*, Energy, Environment, and Sustainability, https://doi.org/10.1007/978-981-96-6628-7_9

9.1 Introduction

The global aim to achieve net-zero carbon emissions by 2050 requires several means to reduce the harmful greenhouse gases (GHG) from various sources. The transportation sector is among the largest producers of greenhouse gases. The uncertain future and the production of large amounts of gaseous, particulate, and GHG exhaust emissions from conventional internal combustion (IC) engine fuels, are the driving forces for the researchers to find effective solutions. Alternative fuels are one of these solutions, which can reduce the consumption of fossil fuel and exhaust emissions from IC engines. Ammonia (NH_3) and Hydrogen (H_2) are examples of carbon-zero alternative fuels that can reduce the concentration of gaseous and particulate pollutants from IC engines. These carbon-free fuels have been tested in their pure forms as well as blended with conventional fossil fuels. Ammonia is an attractive alternative fuel and hydrogen carrier with a high-octane rating and energy hydrogen, that reduces HC (Hydrocarbon), CO (Carbon monooxide), CO_2 (Carbon dioxide), SO_X (Sulphur oxides), and PM (Particulate Matter) emissions (Tutak et al. 2023). Ammonia-powered engines can reduce GHG emissions by one-third as compared to gasoline and diesel-powered engines (Bicer and Dincer 2018). It can be transported and stored easily due to its lower density compared to gasoline and diesel fuel, and already well-established transportation and storage facility (Mounaïm-Rousselle and Brequigny 2020). Ammonia is among the most produced chemicals globally, the current global production of it is around 235 million tonnes (European Maritime Safety Agency 2022). China is the leading producer of ammonia with around 32% of total production in the world (https:ammoniaenergy.org/articles,renewable/ammonia-in-china-full-speed-ahead n.d.). The maximum amount of ammonia production (~80%) is utilized as nitrogen-based fertilizers like ammonium nitrate and urea. The remaining amount is used in other industrial applications such as the production of nitric acid, explosives, plastics, refrigerants, etc., apart from these as alternative fuel and hydrogen carrier. The production and consumption of ammonia will increase exponentially in the future, according to the projections by Erdemir and Dicer (2021), By 2050, the worldwide production of ammonia will increase up to approximately 1.2 billion tonnes, which is about 8.2 times high as compared to the global production in 2019 (Fig. 9.1).

The researchers are interested in utilizing ammonia directly in the IC engine without converting it into hydrogen, because the conversion of ammonia into hydrogen is way costlier. Also, the storage and transportation of hydrogen is very difficult as compared to ammonia (Thomas and Parks 2006). In the liquid state, ammonia can be stored at a temperature of 298 K and pressure of 10.2 atm (Li et al. 2021).

The various advantageous properties of ammonia that make it suitable as an IC engine fuel are as follows;

- High octane number compared to gasoline
- In the liquid state, it contains about 48% more H_2 by volume than H_2
- Easy and safe storage compared to H_2

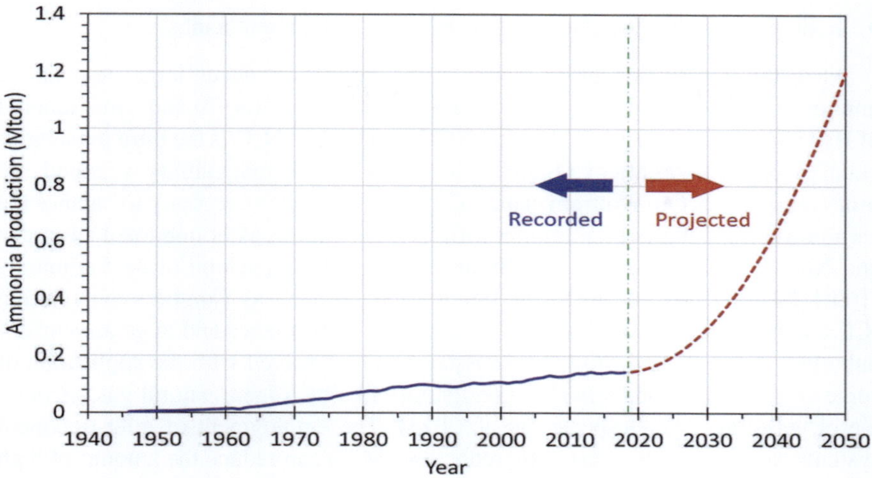

Fig. 9.1 Recorded and projected amount of worldwide ammonia production (Erdemir and Dincer 2021)

- Already established storage and transportation infrastructure
- Can combust in both spark and compression ignition engines
- Ability to reduce carbon and soot emissions.

However, some challenges restrict its utilization as fuel to some extent as follows (Tornatore et al. 2022);

- High auto-ignition temperature (Table 9.1)
- High ignition delay period
- Low laminar flame velocity

Table 9.1 A comparison of different properties of ammonia, gasoline, and diesel (Tornatore et al. 2022; Cai et al. 2023; Wolfram et al. 2022; Jespersen et al. 2024)

Properties	Ammonia	Gasoline	Diesel
Storage method	Compressed liquid	Liquid	Liquid
Density [25 °C, 1 atm]	602.8	698.3	838.8
Lower heating value [MJ/Kg]	18.8	44.5	43.5
Laminar flame velocity [m/s]	0.07	0.58	0.87
Auto-ignition temperature _[°C]	651	300	230
Absolute min. ignition energy [MJ]	8.0	0.14	–
Flammability limits [gas in air vol.%]	15–28	0.6–8	0.6–5.5
Octane rating [RON]	>130	90–98	–
Energy density [MJ/m^3]	11,333	31,074	36,403
Latent heat of vaporization [kJ/kg]	1369	348.7	232.4

- Toxic nature that affects both engine material and human beings.

Also, the combustion of ammonia in IC engines produces large amounts of unburned ammonia and NO_X, particularly Nitrous Oxide (N_2O) and some amount of Hydrogen cyanide (HCN) emissions (Cai et al. 2023). N_2O is the third most influential greenhouse gas after CO_2 and methane, its heat-trapping ability is around 300 times more than CO_2 (Wolfram et al. 2022). Therefore, it is necessary to balance the carbon-free advantages of ammonia with the high emissions of unburned ammonia and NO_X from NH_3 combustion. Studies show that combustion of neat ammonia (100% NH_3) is possible in Spark Ignition (SI) as well as Compression Ignition (CI) engines, but it creates issues of poor engine performance and large amounts of unburned ammonia and NO_X. These issues can be resolved with the application of some combustion promoters or flame ignitors like hydrogen, natural gas, alcohol, diesel, gasoline, etc. (Jespersen et al. 2024). The employment of after-treatment systems, such as selective catalytic reduction (SCR) can reduce the amount of high NO_X and unburned ammonia. Therefore, using ammonia as an IC engine fuel will be more beneficial for stationary or marine engines. In these types of engines, the space limitation for the installation of after-treatment systems is not an issue.

This chapter describes the various production techniques of green ammonia and discusses in detail the methods for utilizing ammonia in SI and CI engines. Finally, the emission characteristics of ammonia-powered SI and CI engines are provided in tabular form with details of the engine type, fuel used, and experimental study results.

9.2 Production Techniques of Green Ammonia

The conventional way of gaseous ammonia production is the Haber–Bosch process, from direct reaction of gaseous H_2 as well as N_2 in the presence of iron, uranium, ruthenium, etc., based catalysts at a temperature between 400 and 450 °C and pressure around 200 atm according to reaction 9.1 (Valera-Medina et al. 2021). The liquid ammonia is received by further cooling and condensation of gaseous ammonia (Tornatore et al. 2022).

$$3H_2 + N_2 \leftrightarrow 2\,NH_3 \qquad \Delta H = -92\ \mathrm{kJ/mol} \tag{9.1}$$

For the synthesis of ammonia, the production of nitrogen on a large scale is carried out by the refinement of liquid air, which is produced with the help of Linde or Claude methods (Agrawal and Woodward 1991). The production of hydrogen for ammonia synthesis is conventionally carried out from natural gas by the steam methane reforming process or by partial oxidation of hydrocarbon, and from coal through coal gasification (Fig. 9.2).

The ammonia produced by utilizing conventional fossil fuels is known as "grey ammonia". Grey ammonia emits CO_2 in the range of 1.51–6.18 tonnes CO_2 per

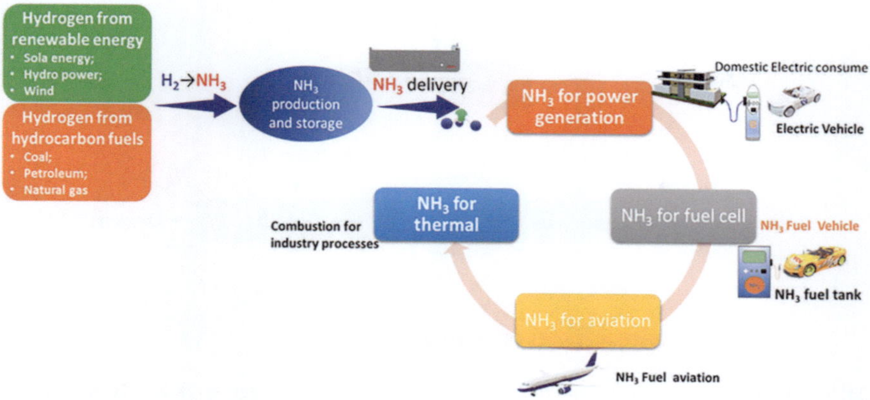

Fig. 9.2 Production and applications of ammonia fuel (Li et al. 2021)

tonne of ammonia produced (OECD, IEA 2021). Ammonia production by natural gas with the involvement of the carbon capture and storage technique is known as "blue ammonia". The production of blue ammonia reduces the total CO_2 emissions by up to 50% as compared to grey ammonia (Saygin et al. 2023). The production of ammonia by utilization of renewable energy sources like solar, hydro, wind, tidal, geothermal, nuclear, biomass, etc. is known as "green ammonia". Green ammonia is almost free of CO_2 emissions because it is produced by using green hydrogen (hydrogen produced by utilizing renewable energy sources) and electricity generated through renewable power sources. The global production of green ammonia in 2021 was only 0.01% of the total ammonia production worldwide. However, it is expected to increase at a rapid rate with production of up to 8% green ammonia by 2030 (Innovation outlook: Renewable ammonia 2022). The major challenge to the way of production of green ammonia is the cost of production associated with it. The production cost of green ammonia at present ranges between USD 400 and 1670 per tonne, depending on the region around the world. Which is almost double the production cost of grey ammonia (Saygin et al. 2023). In 2020, the total production of ammonia was around 185 million tonnes. In which the share of ammonia production through natural gas by steam reforming was around 72%, coal gasification was around 26%, and petroleum oil was around 1% (The Royal Society 2020).

9.2.1 Ammonia Production Through Green Hydrogen + Haber–Bosch Process

In this technique, the green ammonia is produced with the utilization of green hydrogen and the Haber Bosch process combinedly. Water electrolysis is the most frequently used method for the production of green hydrogen (Kumar and Himabindu 2019). In this method, the molecules of water are separated into hydrogen and oxygen

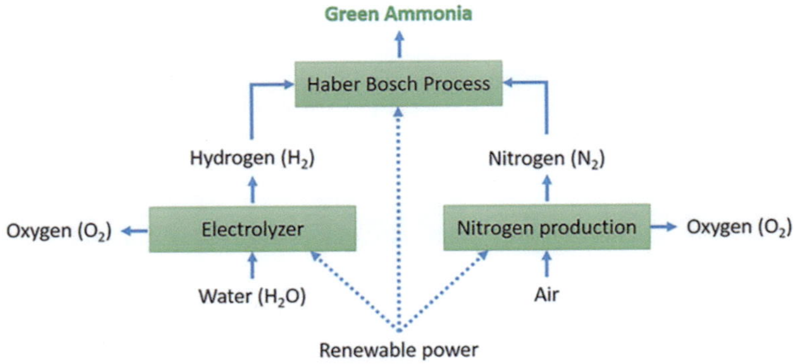

Fig. 9.3 Production of green ammonia through electrolysis of water + Haber–Bosch process (Tornatore et al. 2022)

with the assistance of renewable electricity in an electrolyzer. Nitrogen production is carried out with air separation units, and these air separation units are also run by renewable electricity. In both of these processes, oxygen is obtained as a by-product. The production of per ton of ammonia produces approximately 1.4 tons of oxygen, which can be supplied for medical and industrial purposes after purification (Smith et al. 2022). After the production of green hydrogen (H_2) as well as nitrogen (N_2) in gaseous form, it reacted directly with the production of green ammonia in the presence of a metal catalyst (Fig. 9.3) (Tornatore et al. 2022). For the production of green ammonia with negligible CO_2 emission, all these intermediate processes must be carried out by utilizing the energy generated through renewable sources (Smith and Torrente-Murciano 2021).

The traditional Haber-Bosch process for the production of ammonia is very much energy-intensive and air-polluting. It requires more than 30 GJ of energy, for the production of 1 tonne of ammonia due to the requirement of very high temperature and pressure. Also, it releases about 2.16 tonnes of CO_2 emissions per tonne of ammonia production (Yoo et al. 2013). Whereas, this technique reduces CO_2 emission by around 60–80% compared to fossil fuel-based ammonia production. The range of carbon emission through this technique is as low as 0.12–0.53 tonne of CO_2 per tonne of ammonia production (Ojelade et al. 2023). However, the cost associated with the production of green hydrogen is a critical factor, which increases the cost of green ammonia.

9.2.2 Ammonia Production Through Electrochemical Ammonia Synthesis

Due to various disadvantages of the Haber-Bosch process such as low efficiency, high energy requirement, and a large amount of CO_2 emission, etc., the researchers developed various methods for the production of ammonia without using the Haber-Bosch process (Wu et al. 2021a). Photocatalysis, biocatalysis, and electrocatalysis are the major routes for the production of green ammonia. Electrochemical synthesis of ammonia through catalytic nitrogen reduction by utilizing renewable energy is a more sustainable, energy-efficient, environment-friendly, and scalable process as compared to the Haber-Bosch process (Shen et al. 2021). Generally, the production of ammonia through this process comprises three steps (Wu et al. 2021b):

(i) The adsorption of nitrogen molecules upon the surface of the catalyst
(ii) Breaking the triple covalent bond of nitrogen by an associative or dissociative pathway along with the addition of hydrogen atoms by breaking of H-H bonds,
(iii) Desorption of ammonia, which is generated by the reaction between nitrogen and hydrogen atoms.

The advantages of utilizing the electrochemical technique over the thermochemical (Haber-Bosch) technique are as follows (Soloveichik 2017):

- The efficiency of electrochemical techniques is higher than thermochemical techniques
- It provides higher selectivity, therefore less purification is required
- It requires lower temperatures and pressures, which reduces the capital expenditure
- The modularity of these techniques makes their utilization suitable for small to medium-scale
- Ammonia can be synthesized at ambient conditions.

Despite the various advantages of electrochemical techniques, their low production rates (general range of 10^{-13}–10^{-8} mol s^{-1} cm^{-2}) is a big challenge. This limits the production of ammonia at large scales (Ghavam et al. 2021).

9.3 Methods of Utilizing Ammonia as IC Engine Fuel

Ammonia is a carbon-free alternative fuel; therefore, it does not produce CO_2 and soot emissions during combustion. The research community wanted to utilize these advantageous properties of NH_3 by using it in conventional IC engines as a replacement for fossil fuels. Therefore, many studies have been carried out on conventional SI and CI engines with neat (100%) ammonia. However, due to different physical and chemical properties (Table 9.1) such as lower laminar burning velocity, lower heating value, cetane rating, energy density, etc., and higher auto-ignition temperature, flammability limits, and higher ignition delay, the combustion of neat ammonia is not satisfactory. Also, there was a large amount of unburned ammonia and NO_X emissions.

Therefore, to achieve the proper combustion and a low amount of exhaust emissions, either engine modifications or fuel modifications are required. The modifications in fuel by mixing conventional diesel, gasoline, CNG, or alcohol fuels with

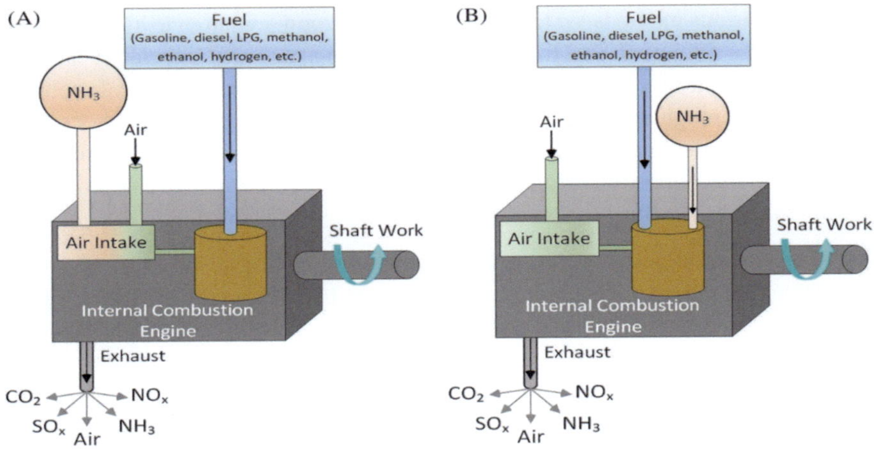

Fig. 9.4 Methods of utilizing ammonia with other fuels in internal combustion engines: **(A)** gaseous ammonia with intake air **(B)** separate injections of liquid ammonia with other fuel (Erdemir and Dincer 2021)

ammonia is an easy way to resolve the issues of high auto-ignition temperature, ignition delay, low laminar flame velocity, etc. There are two commonly used methods for the utilization of ammonia with other fuels (Fig. 9.4) (Erdemir and Dincer 2021).

The first method is to combine the gaseous ammonia with air, after that, the mixture is supplied to the engine cylinder through the intake manifold for air where it ignites easily with the help of other high laminar burning velocity fuels (Port Fuel Injection). The other method is the separate injection of other fuels and liquid ammonia in the engine cylinder with separate injectors (Direct Injection).

9.3.1 *Ammonia as Fuel in SI Engines*

The high-octane rating of ammonia makes it suitable as an SI engine fuel. Also, it reduces greenhouse gas emissions by around one-third compared to gasoline-only operations (Bicer and Dincer 2018). Neat ammonia can be used for SI engine operations at lower compression ratios than CI engines (Starkman et al. 1967). Cornelius et al. (1965) studied the consequence of neat ammonia on the performance and emission of single and multi-cylinder engines. They reported that the engine performance was poor, as well as high emissions of unburned NH_3. However, the low burning velocity, lower heating value, and high autoignition temperature of ammonia create difficulty in operating an SI engine with neat ammonia at low loads. The emission of unburned ammonia is also very high with neat ammonia application. However, there are several ways such as increased compression ratio, supercharging, and dual fuel approach with faster-burning fuel, etc. suggested by the researchers for the efficient operation of SI engines (Xin et al. 2022; Dinesh and Kumar 2022). Due to the lower reactivity of ammonia, the dual fuel approach is very beneficial for SI engines. Different high-reactivity fuels such as gasoline, hydrogen, dimethyl ether (DME), CNG, etc. were combined with ammonia to intensify the laminar burning velocity and decrease the ignition delay and autoignition temperature (Lesmana et al. 2020).

The co-fueling of these fuels also reduces the unburned ammonia emissions. Hydrogen is the most studied fuel among all these, due to the reason that it enhances the engine performance as well as emission characteristics. Also, this hydrogen can be produced by the decomposition of ammonia, due to which the engines can operate with only one fuel tank (Gill et al. 2012; Ryu et al. 2014a). According to the experimental investigation by Morch et al. the mixture of NH_3 and 10% by volume hydrogen gives the best performance in terms of power and thermal efficiency (Mørch et al. 2011). Further, in the study of Oh et al. the combination of 50% NH_3 and 50% CNG reduce CO_2 emissions by about 28% (Oh et al. 2021). Gasoline was also studied as a fuel ignitor for NH_3, a mixture of gasoline and NH_3 helped reduce the NO_X, CO, and HC emissions (Ambalakatte et al. 2023).

In many studies, gaseous ammonia was injected through an intake port (Port Fuel Injection) along with high-reactivity fuel using dedicated injectors. It can be seen in the schematic of the experimental setup of Tutak et al. (2023) (Fig. 9.5) given below.

However, in most recent studies on ammonia, direct injection of ammonia was preferred over port injection. The reason behind this is that the volumetric efficiency

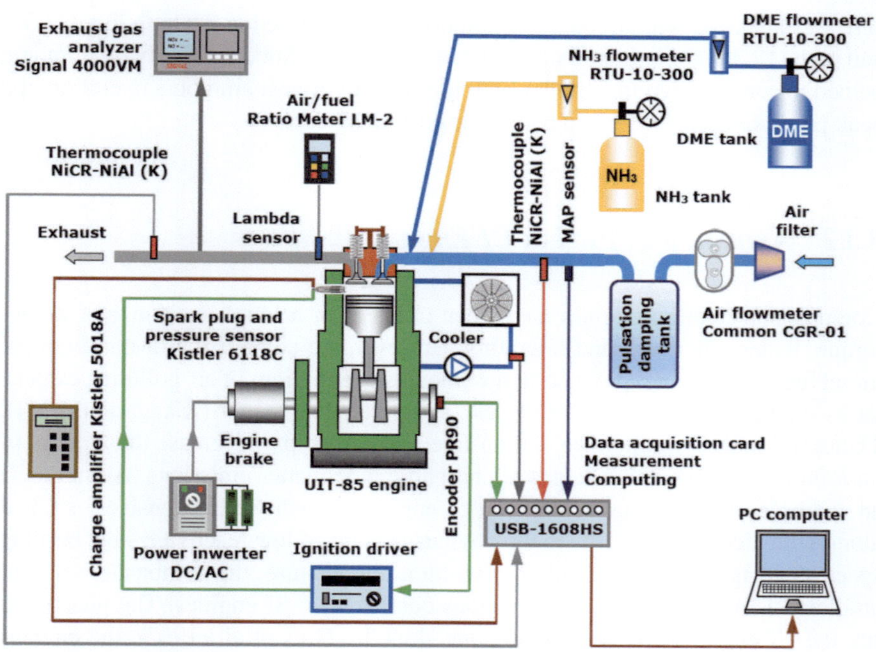

Fig. 9.5 Schematic of the experimental setup where NH_3 and DME were injected through PFI (Tutak et al. 2023)

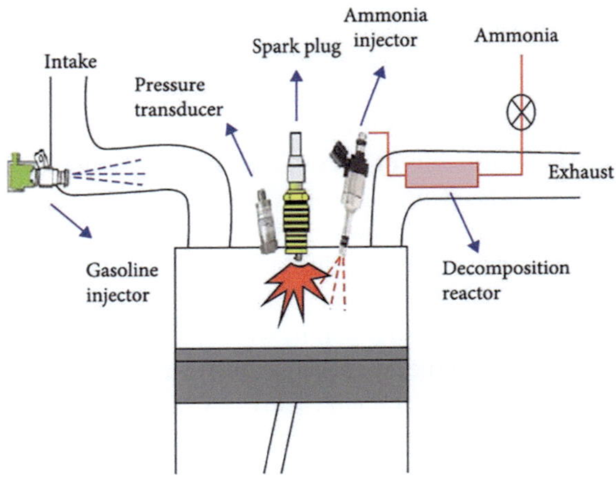

Fig. 9.6 Direct injection of NH_3 with port injection of Gasoline (Ryu et al. 2014a)

of the engine can be increased by direct injection and it also increases the operating range of the engine (Ryu et al. 2014b). Ammonia can be injected directly into the engine cylinder either with port injection of other fuels (Fig. 9.6) or as a mixture of fuels through a single injector (Ryu et al. 2014a).

The effect of ammonia co-fueling on in-cylinder pressure is studied by Ambalakatte et al. (Fig. 9.7) (Ambalakatte et al. 2023). The maximum in-cylinder pressure was achieved with around 60% ammonia energy fraction for both low (11.33) and high (12.39) compression ratios. The excellent anti-knock property of ammonia helped to achieve high in-cylinder pressure. While for neat ammonia operation, the peak pressure decreased compared to the co-fueling case.

9.3.2 *Ammonia as Fuel in CI Engines*

Compression ignition engines have been utilized for a long time for their higher torque, better efficiency, and reliability. However, the combustion of conventional diesel fuel in CI engines produces large amounts of emissions of air-polluting exhaust gases such as NO_X, SO_X, CO CO_2, etc. and particulate matter (Xiang et al. 2024). Therefore, the use of alternate carbon-free fuel can help to decrease the amount of harmful exhaust, greenhouse gases, and particulate matter emissions. Ammonia is an example of an alternative fuel for CI engines, which were extensively studied along with alcohol fuels. Due to the obvious reasons of low reactivity, slow burning speed, high ignition delay, and auto-ignition temperature, the combustion of neat ammonia is more difficult in CI engines compared to SI engines. The attempt to run the CI engine on neat ammonia was done by Gray et al. (1967), the engines ran successfully but for this, they had to increase the compression ratio up to 35:1 at 900 rpm. Conventional CI engines require modifications to the engine design to work at such high compression ratios. The other way to utilize ammonia in a CI

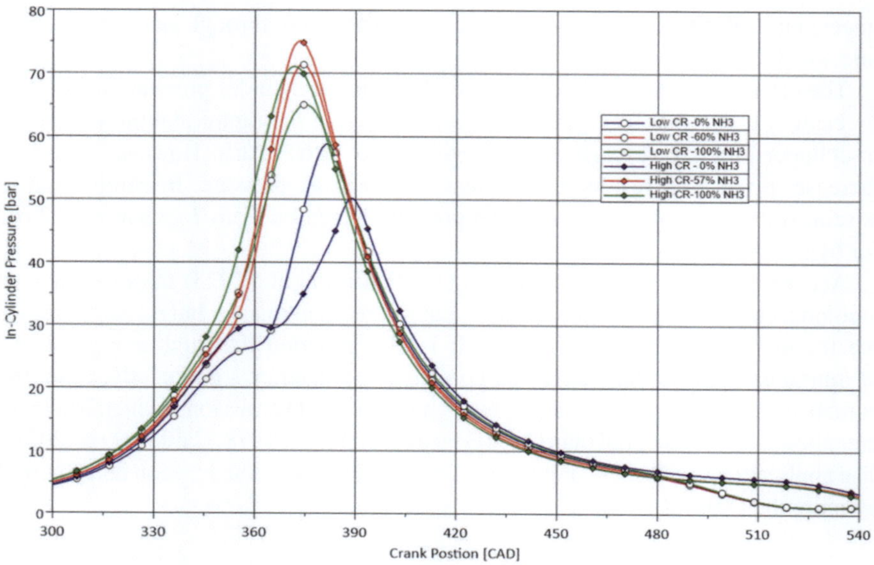

Fig. 9.7 Variation of in-cylinder pressure with ammonia energy fractions and compression ratio (Ambalakatte et al. 2023)

engine is dual dual-fuel approach. With a small amount of high reactivity fuel such as diesel, hydrogen, alcohol fuel, and CNG, etc. as combustion igniter, the engine can be operated with a lesser compression ratio, engine modifications, and less amount of exhaust emissions (Dimitriou and Javaid 2020). The dual fuel approach with direct injection of diesel into engine cylinders and port fuel injection of ammonia (Fig. 9.8) is an effective way for CI engines (Takizawa et al. 2008).

Some researchers have also experimented with direct injection of ammonia as a mixture with other fuels through a single main injector. Ryu et al. (2014c) directly

Fig. 9.8 Direct injection of Diesel with port injection of NH_3 (Tutak et al. 2024)

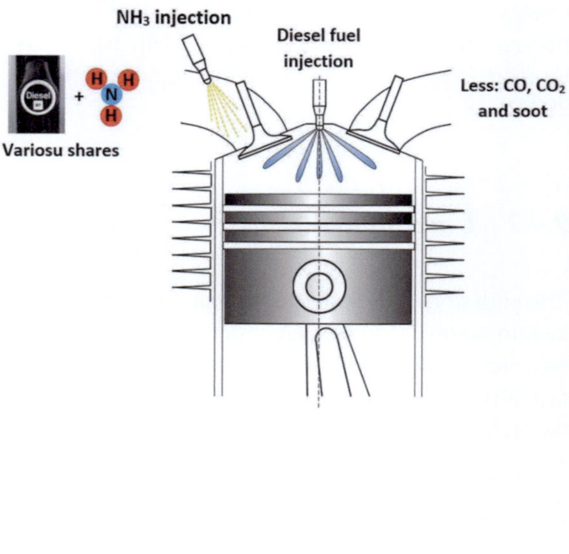

injected the mixture of NH_3 and DME in their experiment through an electronic fuel injector (Fig. 9.9).

The effects of ammonia energy share (AES) on in-cylinder pressure are studied by Nadimi et al. (2023) (Fig. 9.10). It can be seen in the figure that the maximum in-cylinder pressure reached on peak with up to 61.6% AES. However, a further increase in the AES results in a decrease in the peak pressure. In comparison to baseline diesel, the peak of in-cylinder pressure increased from 76.3 bar to 86.2 bar for 61.6% AES.

According to the experimental study by Nadimi et al. (2023) the increment in ammonia energy shares reduces the CO_2 emissions significantly, but it also decreases the thermal efficiency of the engine. It is also reported that high energy shares of ammonia reduced the power and combustion efficiency significantly. Also, the combustion of ammonia produces a high amount of N_2O emissions, which has more greenhouse effect potential than CO_2 (Haar and Gallagher 1978; Cardoso et al. 2021). The application of 5% methanol with ammonia can shorten the ignition delay period by more than 60% (Li et al. 2023)

9.4 Emission Characteristics of Ammonia Powered SI Engines

9.4.1 Regulated Emissions

In most of the studies, the application of around 10% H_2 provided better combustion stability and reduced NO_X emissions (Lhuillier et al. 2019a). However, the NO_2 and NO emissions were increased at high NH_3 dissociation levels (Mercier et al. 2022). A significant reduction in NO_X was also observed, with 18% by volume DME, compared to neat ammonia (Tutak et al. 2023). When ammonia is combined with gasoline, it reduces the NO_X, CO, and HC emissions by increasing the ammonia energy fraction (Figs. 9.11 and 9.12), significantly affecting maximum in-cylinder pressure (Ambalakatte et al. 2023). With 100% NH_3, the N_2O emission decreased by increasing the intake pressure and advancing the spark timing (Jespersen et al. 2024).

9.4.2 Unregulated Emissions

The emission of unburned ammonia from ammonia-fueled engines is a challenging task to resolve. Studies suggested that around 10% by volume H_2 can reduce the emission of unburned ammonia (Mounaïm-Rousselle and Brequigny 2020). With neat NH_3, the emission of unburned NH_3 can be decreased significantly by increasing the brake mean effective pressure (BMEP) (Zhu et al. 2024). A comparative summary

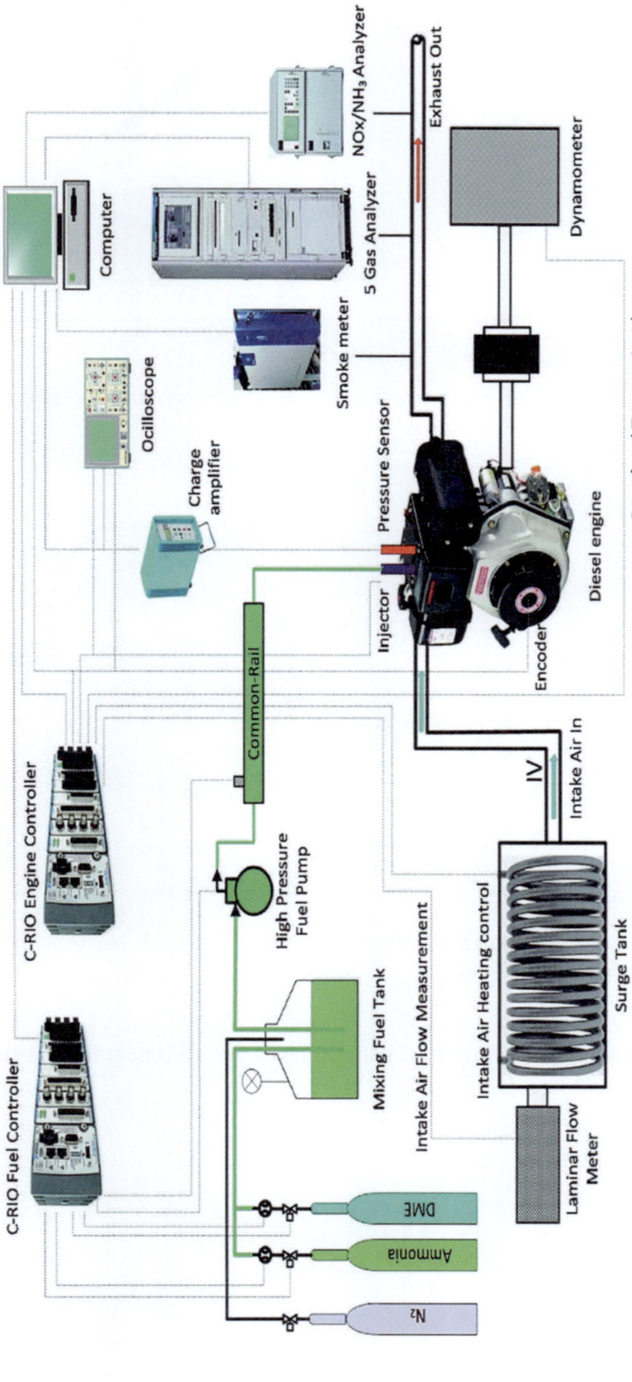

Fig. 9.9 Mixture of ammonia with DME supplied through direct injection Ryu et al. (2014c)

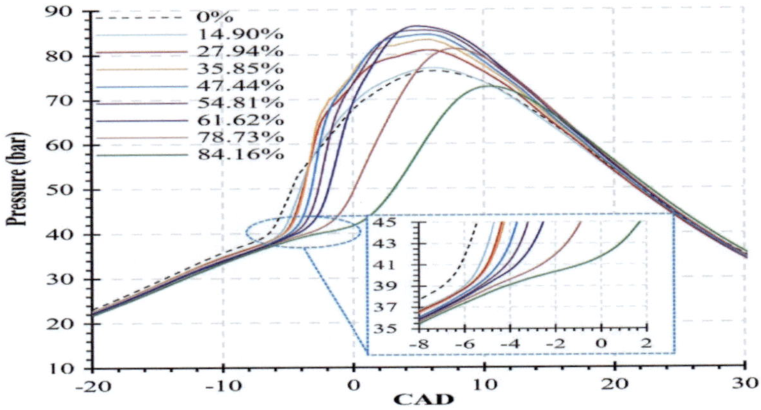

Fig. 9.10 Effect of different ammonia energy shares on in-cylinder pressure (Nadimi et al. 2023)

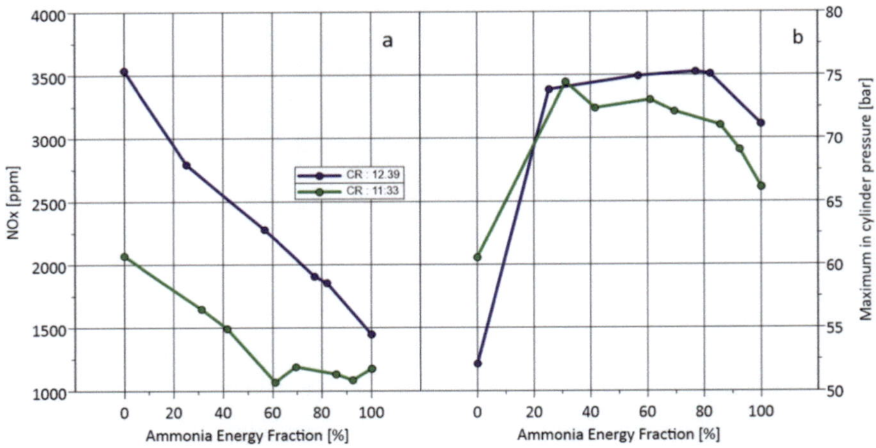

Fig. 9.11 Effect of increasing ammonia fraction on **a** NOx emissions, **b** Maximum in-cylinder pressure (Ambalakatte et al. 2023)

of regulated and unregulated emission characteristics of different ammonia-powered SI engines is provided in Table 9.2.

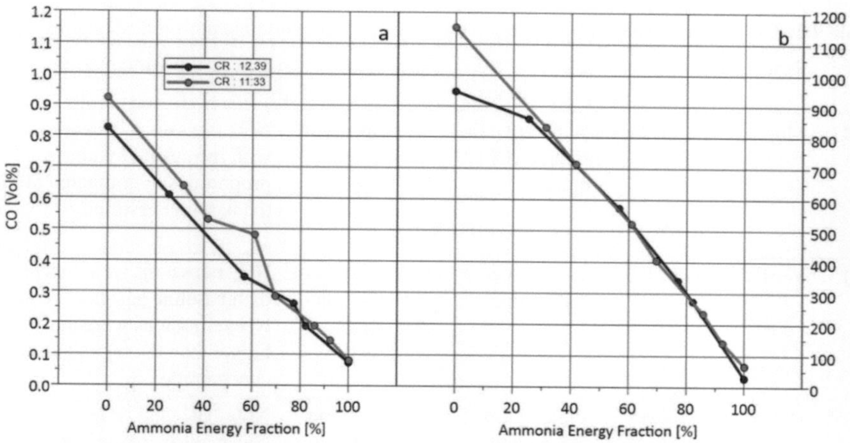

Fig. 9.12 Effect of increasing ammonia fraction on **a** CO emission, **b** THC emissions (Ambalakatte et al. 2023)

9.5 Emission Characteristics of Ammonia Powered CI Engines

9.5.1 Regulated Emissions

The emissions of NO_X were observed to decrease by 61.75% at 60% energy share of aqueous ammonia with diesel in ammonia-fuelled CI engines (Al-Dawody et al. 2023). Beyond that, the NO_X emissions were seen to increase. In another study, CO emission decreased but NO emission increased by increasing the ammonia energy share (AES) with diesel (Fig. 9.13), whereas PM emissions were reduced with the same conditions (Fig. 9.14). CO, CO_2, and HC emissions were decreased by up to 50% through 50% NH_3 energy share with methanol and also through a high amount of biodiesel with NH_3. The utilization of natural gas with NH_3 increased the CO, CO_2, NO_X, and soot emissions, but N_2O emissions decreased significantly by increasing the natural gas proportions (Wang et al. 2024).

9.5.2 Unregulated Emissions

Generally, the emission of unburned ammonia, also known as ammonia slip, increases with increasing ammonia energy shares (Fig. 9.15). It can be reduced by increasing the amount of other fuel energy shares. The emission of Hydrogen cyanide (HCN) is also seen to be increased by more than five times as compared to baseline diesel with about 20% NH3 energy share (Cai et al. 2023), which can be reduced by advancing the main injection timing. A comparative summary of regulated and unregulated

Table 9.2 Experimental studies for emission characteristics of ammonia powered SI engines

Year	Researcher/s	Engine specifications	Fuel used	Test results for emissions
2011	Morch et al. (2011)	Cooperative fuel research (CFR) engine	A composition of NH_3 and H_2, H_2 varied 5–100% by volume	NO_X emissions increased with increasing hydrogen proportions as compared to the baseline gasoline
2013	Westlye et al. (2013)	Single-cylinder CFR engine	Composition of 80% NH_3 and 20% H_2 by volume	NO_2 emissions were higher than baseline gasoline. Also, NH_3 emissions were much higher than the regulated concentrations
2014	Ryu et al. (2014a)	Single-cylinder CFR engine	Port-injection of gasoline to improve the burning of directly injected NH_3	NO_X, HC, and NH_3 emissions increased by increasing the NH_3 injection amount. However, CO emissions were reduced slightly
2019	Lhuillier et al. (2019b)	Single-cylinder, 4-stroke GDI engine	100% NH_3, and composition of 85% NH_3 + 15% H_2	NO_X and NH_3 emissions were reduced with near-stoichiometric lean operation
2019	Lhuillier et al. (2019a)	Single-cylinder, 4-stroke GDI engine	NH_3 with 5–15% by volume H_2	NH_3 with about 10% by volume H_2, provided good combustion efficiency with moderate NH_3 and NO_X emissions
2020	Mounaim-Rousselle and Brequigny (2020)	Four-cylinder, 4-stroke GDI engine	NH_3 with 0%, 5% and 10% by volume H_2	NH_3 with 10% by volume H_2 reduced the NH_3 emissions but increased NO_X and N_2O emissions
2022	Mercier et al. (2022)	Single-cylinder, SI engine	100% NH_3 and Dissociation of NH_3 up to 40%	N_2O, NO, and H_2 emissions were increased at high NH_3 dissociation levels, However, the amount of NH_3 emission was reduced
2023	Ambalakatte et al. (2023)	Single-cylinder, 4-stroke PFI engine	100% NH_3, 100% Gasoline (E10), and a combination of NH_3 and Gasoline	NO_X emissions were reduced by up to 60% but NH_3 slip increased with neat NH_3. Also, CO and HC emissions were decreased by increasing the NH_3 energy fractions
2023	Tutak et al. (2023)	Single-cylinder CFR engine	NH_3 with 0%, 8%, 10%, 13%, 18%, 24%, and 29% by volume DME	NO_X emissions reduced significantly at the optimal value of 18% DME energy share, as compared to neat NH_3

(continued)

Table 9.2 (continued)

Year	Researcher/s	Engine specifications	Fuel used	Test results for emissions
2024	Zhu et al. (2024)	Four-cylinder, 4-valve direct injection engine	NH_3 with 2–16% by volume blending of H_2	With neat NH_3, the emission of unburned NH_3 decreased significantly by increasing the BMEP. Also, by increasing H_2 content the NO_X and unburned NH_3 decreased significantly
2024	Jespersen et al. (2024)	Single-cylinder CFR engine	100% NH_3	The NH_3/NO_X ratio and N_2O emission decreased by increasing the intake pressure. Also, advancing the spark timing provided an NH_3/NO_X ratio of unity and the lowest N_2O emission

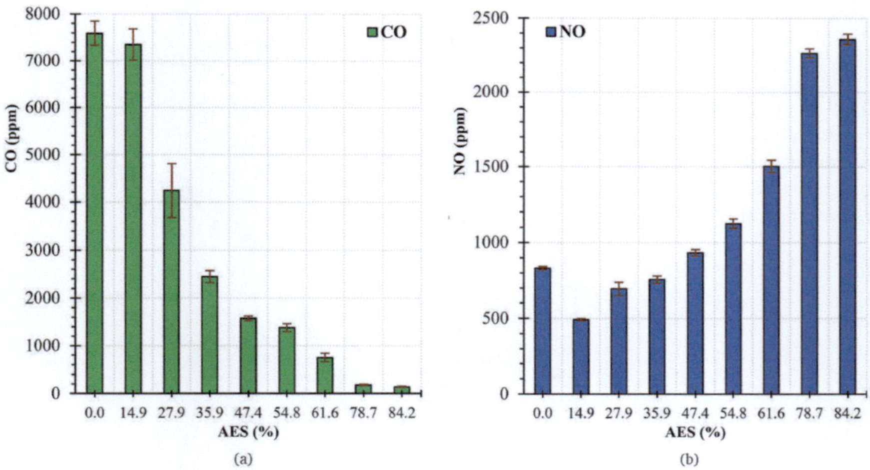

Fig. 9.13 Effects of increasing ammonia energy share **a** on CO emission and **b** on NO emissions (Nadimi et al. 2023)

emission characteristics of different ammonia-powered CI engines is provided in Table 9.3.

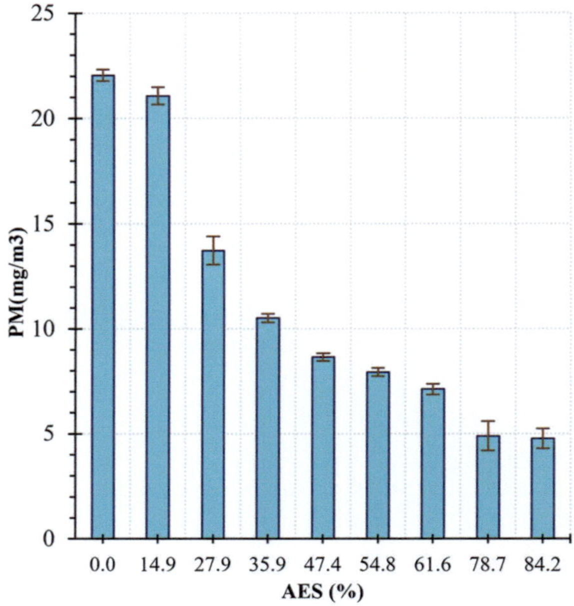

Fig. 9.14 Effects of increasing ammonia energy share on PM emissions (Nadimi et al. 2023)

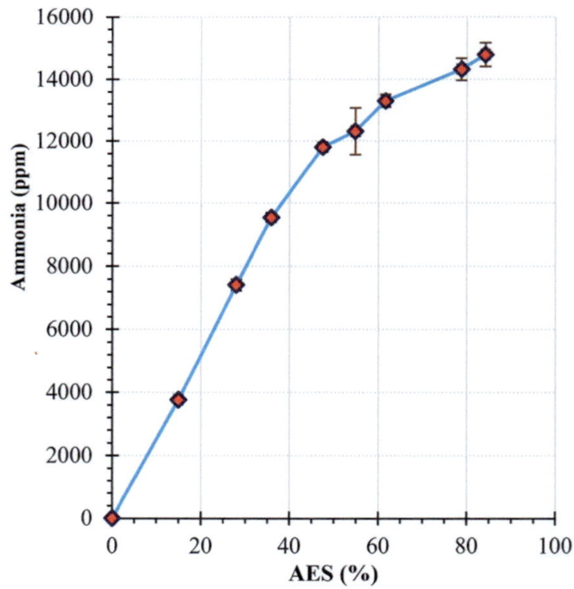

Fig. 9.15 Unburnt ammonia concentrations for various ammonia/diesel ratios in the ammonia-fueled diesel engine (Nadimi et al. 2023)

Table 9.3 Experimental studies for emission characteristics of ammonia powered CI engines

Year	Researcher	Engine specifications	Fuel used	Test results for emissions
2011	Reiter and Kong (2011)	Four-cylinder, turbocharged engine	NH_3 energy fraction of 60–40% with 40–60% energy fraction by Diesel	NO emission increased with more than 60% NH_3 energy share, and CO and HC emissions increased by co-fuelling with NH_3 as compared to baseline diesel. Soot emissions decreased but NH_3 slip increased with high NH_3 energy share
2014	Ryu et al. (2014c)	Single-cylinder, naturally aspirated DI engine	A mixture of NH_3 with DME in 40–60% and 60–40% by weight respectively	NO_X, CO, HC, and NH_3 emissions increased by increasing the weight percentage of NH_3, However, PM emissions were higher for 40% of NH_3 as compared to 60%
2017	Pochet et al. (2017)	Single-cylinder, Homogeneous Charge Compression Ignition (HCCI) engine	Blends of NH_3 up to 60% by volume with H_2	NO_X emissions increased significantly for NH_3-H_2 blends as compared to pure H2 operation
2019	Niki et al. (2019)	Single-cylinder, naturally aspirated engine	Separate injections of NH_3 and Diesel	N_2O and NH_3 emissions were reduced with multiple diesel injections
2022	Nadimi et al. (2022)	Single-cylinder, Lifan air-cooled engine	The engine was first operated with Biodiesel as two different pilot injections, after that NH_3 was injected	HC, CO, and CO_2 emissions were reduced significantly at the highest amount (69.4% energy share) of biodiesel
2023	Al-Dawody et al. (2023)	Single-cylinder, 4-stroke engine	Three different combinations of aqueous Ammonia (NH_4OH) with Diesel as, 40–60%, 50–50%, and 60–40% ratios	NO_X emissions decreased by 61.75% at 60% NH_4OH, also 40–50% NH_4OH can reduce soot emissions effectively as compared to baseline diesel

(continued)

Table 9.3 (continued)

Year	Researcher	Engine specifications	Fuel used	Test results for emissions
2023	Cai et al. (2023)	Single-cylinder, heavy-duty engine	NH_3 about 20% of the total energy share with Diesel split injection	HCN emissions increased by more than five times compared to baseline diesel. THC, CO, N_2O, and HCN emissions were reduced by advancing the main injection timing
2023	Nadimi et al. (2023)	Single-cylinder, Lifan air-cooled engine	NH_3 energy share up to 84.2% with Diesel	CO_2 emission decreased, but NO, N_2O, and NO_2 emissions increased by increasing the NH_3 energy share
2023	Huang et al. (2023)	Six-cylinder, 4-stroke engine	NH_3 mass fraction up to 50% with Methanol	CO and CO_2 emissions were reduced by about 50%, and HC also reduced significantly, but NO_X emissions were not affected by increasing the NH_3 mass fraction
2024	Xu and Bai (2024)	Six-cylinder, heavy-duty DI RCCI engine	NH_3 premixed energy fraction of 0.3–0.80 with Diesel	NO decreased but unburned NH_3 emission increased by increasing the NH_3 energy shares
2024	Tutak et al. (2024)	Single-cylinder, naturally aspirated DI engine	NH_3 energy fraction up to 55% with Diesel	CO and CO_2 emissions decreased, whereas THC first decreased and then increased by increasing the NH_3 fraction. Soot emissions decreased by 3.6 times with 11% NH_3 energy share. NOx emission increased by 30% compared to baseline diesel with up to 44% NH_3 share
2024	Wang et al. (2024)	Single-cylinder, TBD620 engine	NH_3 proportion of 10–100% and Natural Gas proportions of 0–90%, ignited by Diesel	CO, CO_2, NO_X, and soot emissions increased, but N_2O emissions decreased significantly by increasing the natural gas proportions
2024	Xiang et al. (2024)	4-cylinder, turbocharged, CRDI engine	NH_3 fraction of 0%, 10%, 20%, 30% and 40% with Diesel	NO_X emissions were reduced by 43.8%, but GHG emissions increased by 44.8% at 40% NH_3 and 50% load, NH_3 emissions also increased significantly at high NH_3 fraction

9.6 Conclusions

Ammonia emerges as one of the potential carbon-free alternative fuels for IC engines. It can reduce CO, CO_2, THC, SO_X, and PM emissions. The complete combustion of ammonia produces only N_2 and water without the formation of NO_X. However, it produces a relatively higher amount of N_2O and unburned NH_3 emissions due to incomplete combustion in practical IC engines. Generally, ammonia is produced by the traditional Haber–Bosch process, which is an energy-intensive and environmentally polluting process. It generally utilizes fossil fuels such as coal, natural gas, oil, etc. Therefore, to minimize the carbon footprint during ammonia production, by utilizing renewable sources of energy, "green ammonia" can be produced. Green ammonia can be produced by two environment-friendly techniques, (i) Ammonia Production through Green Hydrogen + Haber–Bosch Process, (ii) Ammonia Production through Electrochemical Ammonia Synthesis.

Ammonia can be utilized as a neat fuel (100% NH_3) or as a blend with conventional fossil fuels and other alternative fuels. The combustion of neat ammonia in IC engines requires very high compression ratios due to high autoignition temperature and slow chemical kinetics, which require modifications in the engine design. Also, the combustion of pure ammonia produces large amounts of unburned NH_3 and NO_X emissions. Therefore, the blended form of ammonia with fossil fuels is an optimized way to use it as an IC engine fuel. Many fossil and alternative fuels such as gasoline, diesel, CNG, hydrogen, methanol, DME biodiesel, etc. were studied by various researchers to utilize the carbon-free properties of NH_3 and to reduce the emissions of unburned NH_3 and HCN, etc. There are two general ways to utilize NH_3 with other fuels. The first one is the combination of gaseous NH_3 with intake air supplied through the intake port and the second one is the separate direct injection of liquid NH_3 and other fuel into the engine cylinder. Studies indicate that a blending ratio of 40% ammonia and 60% diesel in a dual-fuel method lowers NO_X emissions compared to pure diesel combustion, also advancing fuel injection timing and multiple fuel injections reduces the N_2O and NH_3 emissions. The blending of 50% ammonia and 50% CNG in CI engines reduces CO_2 emissions by 28%. Whereas around 10% H_2 blended with ammonia, appears as a promising solution for the performance and emission characteristics of SI engines. Therefore, ammonia can be utilized in IC engines as a blend with other fuels as of now, the efficient combustion of pure NH_3 might be possible in upcoming years. The excess NO_X and unburned ammonia can be reduced by the application of after-treatment devices. This chapter provides detailed information about the advantages of utilizing ammonia as IC engine fuel, green ammonia production techniques, methods of utilization in SI and CI engines, and finally the emission characteristics of ammonia-powered SI and CI engines.

Abbreviations

CO_2	Carbon dioxide
CO	Carbon monoxide
HC	Hydrocarbon
NO_X	Nitrogen oxides
SO_X	Sulphur oxides
PM	Particulate Matter
H_2	Hydrogen
HCN	Hydrogen cyanide
NO	Nitrogen monoxide
N_2O	Nitrous oxide
NO_2	Nitrogen dioxide
CNG	Compressed Natural Gas
DME	Dimethyl Ether
IC	Internal Combustion
SI	Spark Ignition
CI	Compression Ignition
DI	Direct Injection
CRDI	Common Rail Direct Injection
CFR	Cooperative Fuel Research
PFI	Port Fuel Injection
GDI	Gasoline Direct Injection
RCCI	Reactivity Controlled Compression Ignition
HCCI	Homogeneous Charge Compression Ignition
AES	Ammonia Energy Share

References

Agrawal R, Woodward DW (1991) Efficient cryogenic nitrogen generators: an exergy analysis. Gas Se Purif 5(3):139–150. https://doi.org/10.1016/0950-4214(91)80012-t

Al-Dawody MF, Al-Obaidi W, Aboud ED, Abdulwahid MA, Al-Farhany K, Jamshed W, Eid MR, Raizah Z, Iqbal A (2023) Mechanical engineering advantages of a dual fuel diesel engine powered by diesel and aqueous ammonia blends. Fuel 346:128398

Ambalakatte A, Geng S, Cairns A, Harrington A, Hall J, Bassett M (2023) Evaluation of ammonia-gasoline co-combustion in a modern spark ignition research engine. Carbon Neutrality 2(1):35. https://doi.org/10.1007/s43979-023-00075-3

Bicer Y, Dincer I (2018) Life cycle assessment of ammonia utilization in city transportation and power generation. J Clean Prod 170:1594–1601. https://doi.org/10.1016/j.jclepro.2017.09.243

Cai K, Liu Y, Chen Q, Qi Y, Li L, Wang Z (2023) Combustion behaviours and unregular emission characteristics in an ammonia–diesel engine. Energies 16:7004. https://doi.org/10.3390/en16197004

Cardoso JS, Silva V, Rocha RC, Hall MJ, Costa M, Eusebio D (2021) Ammonia as an energy vector: current and future prospects for low-carbon fuel applications in internal combustion engines. J Clean Prod 296

Cornelius W, Huellmantel LW, Mitchell HR (1965) Ammonia as an engine fuel. SAE Technical Paper 1965:650052. https://doi.org/10.4271/650052

Dimitriou P, Javaid R (2020) A review of ammonia as a compression ignition engine fuel. Int J Hydrogen Energy 45(11):7098–7118

Dinesh MH, Kumar GN (2022) Effects of compression and mixing ratio on NH_3/H_2 fueled SI engine performance, combustion stability, and emission. Energy Convers Manag: X 15:100269. https://doi.org/10.1016/J.ECMX.2022.100269

Erdemir D, Dincer I (2021) A perspective on the use of ammonia as a clean fuel: challenges and solutions. Int J Energy Res 45:4827–4834. https://doi.org/10.1002/er.6232

European Maritime Safety Agency (2022) Update on potential of biofuels in shipping

Ghavam S, Vahdati M, Wilson IAG, Styring P (2021) Sustainable ammonia production processes. Front Energy Res 9:580808. https://doi.org/10.3389/fenrg.2021.580808

Gill S, Chatha G, Tsolakis A, Golunski SE, York AJIjohe (2012) Assessing the effects of partially decarbonising a diesel engine by co-fuelling with dissociated ammonia 37:6074–6083

Gray JT Jr, Dimitroff E, Meckel NT, Quillian RD Jr (1967) Ammonia fuel—engine compatibility and combustion. SAE Trans 1:785–807. https://doi.org/10.4271/660156

Haar L, Gallagher JS (1978) Thermodynamic properties of ammonia. J Phys Chem Ref Data 7(3):635–792

https://ammoniaenergy.org/articles/renewable-ammonia-in-china-full-speed-ahead

Huang Z, Lyu Z, Luo P, Zhang G, Ying W, Chen A, Xiao H (2023) Effects of methanol–ammonia blending ratio on performance and emission characteristics of a compression ignition engine. J Mar Sci Eng 11(12):2388

Innovation outlook: Renewable ammonia (2022) International renewable energy agency and ammonia energy association

Jespersen MC, Rasmussen TO, Ivarsson A (2024) Widening the operation limits of a SI engine running on neat ammonia. Fuel 358:130159. https://doi.org/10.1016/j.fuel.2023.130159

Kumar SS, Himabindu V (2019) Hydrogen production by PEM water electrolysis–a review. Mater Sci Energy Technol 2(3):442–454. https://doi.org/10.1016/j.mset.2019.03.002

Lesmana H, Zhu M, Zhang Z, Gao J, Wu J, Zhang DJF (2020) Experimental and kinetic modelling studies of laminar flame speed in mixtures of partially dissociated NH_3 in air 278:118428

Lhuillier C, Brequigny P, Contino F, Mounaïm-Rousselle C (2019a) Performance and emissions of an ammonia-fueled si engine with hydrogen enrichment. SAE Technical Paper. https://doi.org/10.4271/2019-24-0137

Lhuillier C, Brequigny P, Contino F, Mounaïm-Rousselle C (2019b) Combustion characteristics of ammonia in a modern spark-ignition engine. SAE Technical Paper. https://doi.org/10.4271/2019-24-0237

Li J, Lai S, Chen D, Wu R, Kobayashi N, Deng L, Huang H (2021) A Review on combustion characteristics of ammonia as a carbon-free fuel. Front Energy Res 9:760356. https://doi.org/10.3389/fenrg.2021.760356

Li X, Ma Z, Jin Y, Wang X, Xi Z, Hu S, Chu X (2023) Effect of methanol blending on the high-temperature auto-ignition of ammonia: an experimental and modeling study. Fuel 339:126911

Mercier A, Mounaïm-Rousselle C, Brequigny P, Bouriot J, Dumand C (2022) Improvement of SI engine combustion with ammonia as fuel: effect of ammonia dissociation prior to combustion. Fuel Commun 11:100058

Mørch CS, Bjerre A, Gøttrup MP, Sorenson SC, Schramm J (2011) Ammonia/hydrogen mixtures in an SI-engine: engine performance and analysis of a proposed fuel system. Fuel 90:854–864. https://doi.org/10.1016/j.fuel.2010.09.042

Mounaïm-Rousselle C, Brequigny P (2020) Ammonia as fuel for low-carbon spark-ignition engines of tomorrow's passenger cars. Front Mech Eng 6:70. https://doi.org/10.3389/fmech.2020.00070

Nadimi E, Przybyla G, Emberson D, Løvås T, Ziolkowski L, Adamczyk W (2022) Effects of using ammonia as a primary fuel on engine performance and emissions in an ammonia/biodiesel dual-fuel CI engine. Int J Energy Res 10(1002):8235

Nadimi E, Przybyła G, Lewandowski MT, Adamczyk W (2023) Effects of ammonia on combustion, emissions, and performance of the ammonia/diesel dual-fuel compression ignition engine. J Energy Inst 107:101158

Niki Y, Nitta Y, Sekiguchi H, Hirata K (2019) Diesel fuel multiple injection effects on emission characteristics of diesel engine mixed ammonia gas into intake airJ Eng Gas Turbines Power 141. https://doi.org/10.1115/1.4042507

OECD/IEA (2021) Ammonia technology roadmap. International energy agency. Paris, France

Oh S, Park C, Kim S, Kim Y, Choi Y, Kim C (2021) Natural gas—ammonia dual-fuel combustion in spark-ignited engine with various air-fuel ratios and split ratios of ammonia under part load condition. Fuel 290:120095. https://doi.org/10.1016/j.fuel.2020.120095

Ojelade OA, Zaman SF, Ni BJ (2023) Green ammonia production technologies: a review of practical progress. J Environ Manag 342:118348

Pochet M, Truedsson I, Foucher F, Jeanmart H, Contino F (2017) Ammonia-hydrogen blends in homogeneous-charge compression-ignition engine. SAE Technical Paper 24:0087. https://doi.org/10.4271/2017-24-0087

Reiter A, Kong SC (2011) Combustion and emissions characteristics of compression-ignition engine using dual ammonia-diesel fuel. Fuel 90:87–97. https://doi.org/10.1016/j.fuel.2010.07.055

Ryu K, Zacharakis-Jutz GE, Kong SC (2014a) Performance enhancement of ammonia-fueled engine by using dissociation catalyst for hydrogen generation. Int J Hydrogen Energy 39(5):2390–2398

Ryu K, Zacharakis-Jutz GE, Kong SC (2014b) Effects of gaseous ammonia direct injection on performance characteristics of a spark-ignition engine. Appl Energy. https://doi.org/10.1016/j.apenergy.2013.11.06740

Ryu K, Zacharakis-Jutz GE, Kong SC (2014c) Performance characteristics of the compression-ignition engine using a high concentration of ammonia mixed with dimethyl ether. Appl Energy 113:488–499

Saygin D, Blanco H, Boshell F, Cordonnier J, Rouwenhorst K, Lathwal P, Gielen D (2023) The role of green ammonia in transitioning the non-energy use of natural gas. Sustainability 15:1623. https://doi.org/10.3390/su15021623

Shen H, Choi C, Masa J, Li X, Qiu J, Jung Y, Sun Z (2021) Electrochemical ammonia synthesis: mechanistic understanding and catalyst design. Chem 7(7):1708–1754. https://doi.org/10.1016/j.chempr.2021.01.009

Smith C, Torrente-Murciano L (2021) Guidance for targeted development of ammonia synthesis catalysts from a holistic process approach. Chem Catal 1(6):1163–1172. https://doi.org/10.1016/j.checat.2021.09.015

Smith C, Hill AK, Torrento-Murciano L (2022) Current and future role of Haber-Bosch ammonia in a carbon-free energy landscape. Energy Environ Sci 13:331–344

Soloveichik G (2017) Future of ammonia production: improvement of Haber-Bosch process or electrochemical synthesis. In: The 14th annual NH3 fuel conference

Starkman ES, Newhall HK, Sutton R, Maguire T, Farbar L (1967) Ammonia as a spark ignition engine fuel: Theory and application. SAE Trans 75:765–784. http://www.jstor.org/stable/44563674

Takizawa K, Takahashi A, Tokuhashi K, Kondo S, Sekiya A (2008) Burning velocity measurements of nitrogen-containing compounds. J Hazard Mater 155(1):144–152. https://doi.org/10.1016/j.jhazmat.2007.11.089

The Royal Society (2020) Ammonia: zero-carbon fertiliser, fuel and energy. https://royalsociety.org/-/media/policy/projects/green-ammonia/green-ammonia-policybriefing.pdf. Accessed 15 May 2022

Thomas G, Parks G (2006) Potential roles of ammonia in a hydrogen economy—a study of issues related to the used ammonia for on-board vehicular hydrogen storage. U.S. Department of Energy

Tornatore C, Marchitto L, Sabia P, De Joannon M (2022) Ammonia as green fuel in internal combustion engines: state-of-the-art and future perspectives. Front Mech Eng 8:944201. https://doi.org/10.3389/fmech.2022.944201

Tutak W, Pyrc M, Gruca M, Jamrozik A (2023) Ammonia combustion in a spark-ignition engine supported with Dimethyl Ether. Energies 16(21):7283. https://doi.org/10.3390/en16217283

Tutak W, Jamrozik A, Grab-Rogaliński K, Pyrc M (2024) Effects of ammonia energy fraction on combustion stability and emissions characteristics of naturally aspired industrial dual-fuel diesel engine. Energy Convers Manag 314:118698

Valera-Medina A, Amer-Hatem F, Azad AK, Dedoussi IC, De Joannon M, Fernandes RX, Glarborg P, Hashemi H, He X, Mashruk S, McGowan J (2021) Review on ammonia as a potential fuel: from synthesis to economics. Energy Fuels 35(9):6964–7029. https://doi.org/10.1021/acs.energyfuels.0c03685

Wang B, Yang C, Wang H, Hu D, Duan B, Wang Y (2024) Effects of combustion and emission performance of ammonia/natural gas engines ignited by diesel. Fuel 358:130323. https://doi.org/10.1016/j.fuel.2023.130323

Westlye FR, Ivarsson A, Schramm J (2013) Experimental investigation of nitrogen-based emissions from an ammonia fueled SI-engine. Fuel 111:239–247. https://doi.org/10.1016/j.fuel.2013.03.055

Wolfram P, Kyle P, Zhang X, Gkantonas S, Smith S (2022) Using ammonia as a shipping fuel could disturb the nitrogen cycle. Nat Energy 7:1112–1114

Wu ZY, Karamad M, Yong X, Huang Q, Cullen DA, Zhu P, Xia C, Xiao Q, Shakouri M, Chen FY, Kim JY (2021a) Electrochemical ammonia synthesis via nitrate reduction on Fe single atom catalyst. Nat Commun 12(1):2870. https://doi.org/10.1038/s41467-021-23115-x

Wu T, Fan W, Zhang Y, Zhang F (2021b) Electrochemical synthesis of ammonia: progress and challenges. Mater Today Phys 16:100310

Xiang P, Liu J, Zhao W, Ji Q, Ao C, Wang X, Sun P, Wang X, Li Z (2024) Experimental investigation on gas emission characteristics of ammonia/diesel dual-fuel engine equipped with DOC+SCR aftertreatment. Fuel 359:130496

Xin G, Ji C, Wang S, Meng H, Chang K, Yang J (2022) Effect of ammonia addition on combustion and emission characteristics of hydrogen-fueled engine under lean-burn condition. Int J Hydrogen Energy 47(16):9762–9774. https://doi.org/10.1016/J.IJHYDENE.2022.01.027

Xu L, Bai XS (2024) Numerical investigation of engine performance and emission characteristics of an Ammonia/Hydrogen/n-Heptane engine under RCCI operating conditions. Flow Turbulence Combust 112:957–974. https://doi.org/10.1007/s10494-023-00453-y

Yoo CY, Joo JH, Jang SY, Yu JH, Jeong HN, Hyeong CH, Yoon HC, Kim JN (2013) Electrochemical ammonia synthesis from water and nitrogen using solid-state ion conductors. In: NH3 fuel conference ammonia energy association Washington DC USA

Zhu T, Yan X, Gao Z, Qiu Y, Zhu L, Huang Z (2024) Combustion and emission characteristics of ammonia-hydrogen fueled SI engine with high compression ratio. Int J Hydrogen Energy 62:579–590

Chapter 10
Identification of Critical Equipment and System-Based Risk Factors of an Ammonia-Fueled Marine Diesel Engine

Burak Zincir

Abstract Maritime transport accounts for 90% of global trade. Recent data indicates that maritime transport grew by 2.4% in 2023 and is projected to grow at an annual rate of 2.1% over the next 5 years. The global maritime fleet comprises 105,493 ships above 100 gross tonnages, the majority of which use low-quality fossil-based fuels. Heavy fuel oil constitutes 72% of these fuels, marine diesel oil 26%, and only 2% is a low-carbon fossil fuel, specifically LNG. Along with the emission studies of the International Maritime Organization (IMO), there are considerable low or zero-carbon alternative fuels for maritime transport beyond LNG. One promising alternative is ammonia, which boasts a zero-carbon structure, easy storage capabilities, and an energy density comparable to other alternative fuels like methanol. Utilizing ammonia could help achieve the IMO's 2050 net-zero target. However, ammonia's unique properties must be carefully managed. As a toxic and corrosive substance, ammonia poses significant risks to the marine ecosystem if spilled. Additionally, being a flammable gas, it carries flash, explosion, and fire risks. If stored at −33 °C as a cold liquid, ammonia also presents a cold-burn hazard in the event of a fuel system leak. This chapter identifies critical equipment and system-based risk factors associated with ammonia fuel supply systems and ammonia-fueled marine diesel engines. It highlights key equipment and outlines potential risk factors, including toxicity, corrosiveness, stress corrosion, low temperature embrittlement, BLEVE, and flammability.

Keywords Ammonia · Marine engine · Critical equipment · Fuel system · Risk factors

---

B. Zincir (✉)
Marine Engineering Department, Maritime Faculty, Istanbul Technical University, Istanbul, Türkiye
e-mail: bzincir@itu.edu.tr

© The Author(s), under exclusive license to Springer Nature Singapore Pte Ltd. 2026

H. Valera and A. K. Agarwal (eds.), *Ammonia and Low Carbon Combustion Engine Technologies for De-Fossilizing Transport Sector*, Energy, Environment, and Sustainability, https://doi.org/10.1007/978-981-96-6628-7_10

Nomenclature

AEGL	Acute Exposure Guideline Levels
BLEVE	Boiling Liquid Expanding Vapor Explosion
CII	Carbon Intensity Indicator
DCS	Data Collection System
EEDI	Energy Efficiency Design Index
EEXI	Energy Efficiency Existing Ship Index
ERC	Emergency Release Coupling
FAME	Fatty Acid Methyl Ester
FPSS	Fuel Preparation and Supply System
HTL	Hydrothermal Liquefaction
HVO	Hydrotreated Vegetable Oil
IGF	The International Code of Safety for Ship Using Gases and Other Low-flashpoint Fuels
IMO	International Maritime Organization
LHV	Lower Heating Value
SEEMP	Ship Energy Efficiency Management Plan
UCO	Used Cooking Oil

10.1 Introduction

Maritime transportation constitutes a significant part of global trade. According to recent data, it grew by 2.4% in 2023 and is expected to grow annually by 2.1% over the next 5 years. The global maritime trade fleet consists of 105,493 ships with over 100 gross tons (United Nations Conference on Trade and Development (UNCTAD) 2023). Most of these ships use low-quality, fossil-based fuels, with 72% using heavy fuel oil (HFO) and 26% using marine diesel oil (MDO), while only 2% use LNG (liquefied natural gas), which is also fossil based but has a lower carbon content (Zincir 2022). Due to the use of low-quality fossil-based fuels, global maritime transport accounts for 0.61% of global carbon monoxide (CO) emissions, 9.84% of sulfur oxide (SO_X) emissions, 14.74% of nitrogen oxide (NO_X) emissions, 6.75% of particulate matter 2.5 (PM2.5) emissions, 3.56% of particulate matter 10 (PM10) emissions (European Energy Agency (EEA) 2021), and 2.89% of global carbon dioxide (CO_2) emissions (International Maritime Organization (IMO) 2020).

The International Maritime Organization (IMO) has implemented numerous emission regulations from the past to the present to reduce and control ship emissions. Emission limits for NO_X, SO_X, and PM emissions have been significantly reduced. In addition to these limits, work has also been done on CO_2 emissions and other greenhouse gases (GHGs), and these efforts are ongoing. In 2012, the Ship Energy Efficiency Management Plan (SEEMP) was introduced, focusing on reducing greenhouse gas (GHG) emissions. SEEMP includes indicators like the Energy Efficiency

Design Index (EEDI), Energy Efficiency Existing Ship Index (EEXI), and Carbon Intensity Indicator (CII) to measure carbon emissions. The IMO Data Collection System (DCS) was established to monitor fuel oil consumption. Since January 1, 2023, ships have been assessed based on EEDI, EEXI, and CII, and those failing to meet the criteria cannot be certified (International Maritime Organization (IMO) 2022). The most important of these efforts is the IMO Greenhouse Gas Strategy. With this strategy, maritime transport has, for the first time, aligned with the same goals as the Paris Climate Change Agreement. According to the IMO's greenhouse gas strategy, a 40% reduction in CO_2 emissions compared to 2008 levels by 2030 and net-zero GHG emissions by 2050 have been targeted (International Maritime Organization (IMO) 2023). Measures that can be implemented to achieve these targets are also outlined in this strategy. Looking at the short-, medium-, and long-term measures, the use of low- or zero-carbon alternative fuels in ships stands out as crucial for reaching the IMO's targets. Although 92.6% of the tonnage in operation uses conventional fuels, half the tonnage on order will have alternative-fueled engines (DNV 2024). LNG, which already has a low carbon content, is currently used in maritime transport. In addition to this fuel, low-carbon fuels such as methanol, LPG (liquefied petroleum gas), and ethane are also used and studied in global maritime transport. Similarly, biofuels, produced from carbon-containing products or waste, can be used as alternative fuels in this regard. Apart from carbon-containing fuels, hydrogen and ammonia attract the maritime sector's attention due to their carbon-free nature. Ammonia is particularly prominent for medium and long voyages due to its easier storage properties compared to hydrogen and higher energy density per unit volume. Studies on ammonia-fueled main engines, diesel generators, and their associated fuel systems are ongoing, and they are not yet commercially available. However, engine manufacturers dominating the maritime sector, such as MAN and Wartsila, are producing ammonia-fueled engines and fuel systems. Table 10.1 shows the developments in ammonia fuel in the maritime industry in 2024, and Table 10.2 lists the work of IMO on ammonia as a marine fuel.

Despite increased attention on the prominent alternative marine fuel, ammonia, it has a toxic and corrosive nature. This chapter aims to focus on the properties of ammonia by comparing the properties with other promising alternative marine fuels, ammonia-fueled marine diesel engine system, and critical equipment and system-based risk factors of an ammonia-fueled marine diesel engine.

10.2 Properties of the Alternative Fuels in Maritime Transport

This section consists of the properties of the alternative fuels, LNG, methanol, biofuels, hydrogen, and ammonia. These fuels are selected since they are the most promising alternative fuels in maritime transport (American Bureau of Shipping (ABS) 2024).

Table 10.1 Developments in ammonia fuel in the maritime industry in 2024

Date	Development	Reference
6 June 2024	Lloyd's Register (LR) has approved a 360,000 DWT ammonia-fueled large ore carrier designed in China	Lloyd's Register (LR) (2024a)
7 June 2024	Lloyd's Register (LR) has approved SDARI's design for an 8,200 TEU ammonia-fueled container ship. Developed as part of a collaborative project with Mediterranean Shipping Co and MAN E&S	Lloyd's Register (LR) (2024b)
3 June 2024	Lloyd's Register (LR), Samsung Heavy Industries (SHI), and Amogy have approved the design of an 88,000 cbm ammonia carrier that uses Amogy's ammonia-cracking technology to produce hydrogen and power a fuel cell	Lloyd's Register (LR) (2024c)
21 June 2024	Lloyd's Register (LR) has granted an Approval in Principle (AiP) for HD Korea Shipbuilding & Offshore Engineering's (HD KSOE) ammonia fuel system design. The new system is compatible with high-efficiency cargo handling systems and ammonia engines, and has undergone LR's risk-based certification process	Lloyd's Register (LR) (2024d)
26 June 2024	Norwegian Enova has provided new funding for ammonia-powered vessels. With this support, Amon Maritime has placed orders for two ammonia-fueled gas carriers. Höegh Autoliners will also use the fund to construct ammonia-powered Aurora Class vehicle carriers	Atchison (2024a)
4 July 2024	Norwegian state-owned Enova has allocated $112.7 million in funding to support the construction of 15 hydrogen and ammonia-fueled ships	Riviera (2024)
5 July 2024	Norwegian Eidesvik Offshore received significant approval for the installation of an ammonia fuel system and a 2 MW solid oxide fuel cell on the Viking Energy vessel. The Norwegian Maritime Authority approved the feasibility of this design	TradeWinds (2024)
9 July 2024	Wärtsilä will supply ammonia fuel systems for six vessels being built at Hyundai Mipo Shipyard, contributing to the reduction of emissions in maritime transport. Additionally, they are developing an ammonia-fueled bunker vessel in collaboration with Gas and Heat in Italy. Amogy and Mitsubishi are also collaborating on ammonia and hydrogen fuel systems	Atchison (2024b)
31 July 2024	ClassNK granted the world's first Approval in Principle (AiP) for the bunkering boom design for ammonia fuel developed by Nippon Yusen Kabushiki Kaisha (NYK Line) and TB Global Technologies Ltd	ClassNK (2024)
16 August 2024	In preparation for the increasing ammonia imports at the Port of Rotterdam, the Netherlands' PGS-12 ammonia storage code has been updated. The Port of Rotterdam will play a key role in the import of low-emission energy, and new ammonia terminals will be established	Rouwenhorst (2024)

Table 10.2 International Maritime Organization (IMO) developments on ammonia as a marine fuel (Jang et al. 2024)

IMO meeting	Document title
MSC 104/15/9	Development of non-mandatory guidelines for the safety of ships using ammonia as fuel
MSC 104/15/10	Hazard identification of ships using ammonia as a fuel
MSC 104/15/30	Necessity of deliberations on operational safety measures and fire safety measures
CCC 7/3/9	Report from the correspondence group and proposal for developing guidelines for the use of ammonia and hydrogen as a fuel
CCC 7/INF.8	Forecasting the alternative marine fuel: ammonia
CCC 8/13/1	Development of guidelines for the safety of ships using ammonia as fuel
CCC 8/13/2	Comments on document CCC 8/13
CCC 8/13	Report of the correspondence group (safety information for the use of ammonia)
CCC 9	Amendments to the IGF code and development of guidelines for alternative fuels and related technologies (report of the working group)

10.2.1 Liquefied Natural Gas (LNG)

Liquefied natural gas (LNG) is the liquefied form of natural gas, achieved by cooling it to approximately −162 °C (Elgohary et al. 2014). Natural gas primarily consists of methane, with smaller amounts of ethane, propane, and butane (American Bureau of Shipping (ABS) 2015). Natural gas is considered a cleaner fossil fuel due to its negligible sulfur or ash content, low levels of NO_X, and reduced CO_2 emissions. While natural gas is not flammable in its liquid phase, it becomes highly flammable and ready to burn when present in concentrations ranging from 5 to 15 percent by volume in the air (American Bureau of Shipping (ABS) 2015). It can also be utilized in dual-fuel engines alongside other fuels (Bilgili 2021). The appeal of LNG as a marine fuel has grown due to recent international regulations, advancing technology, and economic factors in the maritime industry. LNG offers a 25% reduction in CO_2 emissions compared to HFO (Iannaccone et al. 2020; Karatuğ et al. 2023). Moreover, LNG is competitively priced relative to other alternative marine fuels.

However, LNG presents certain challenges, such as the need for storage in highly insulated tanks and the requirement for more storage space, potentially leading to additional costs. Another disadvantage is that LNG alone cannot meet the international target of a 50% CO_2 reduction (Gl 2019).

10.2.2 Methanol

Methanol is one of the top five produced chemicals globally (Independent Commodity Intelligence Services (ICIS) 2017), with an annual production capacity of about 95 million tons (Nash 2015), including 20 million tons produced as fuel or fuel blend (Landälv 2017). Methanol can be derived from fossil fuel sources, primarily natural gas and coal, or renewable sources, including wood, agricultural, and municipal waste (Yao et al. 2017). This type of methanol is known as bio-methanol, which significantly reduces GHG emissions compared to fossil fuel-derived methanol (Maritime Knowledge Center (MKC), TNO, TU Delft 2018). Additionally, methanol produced using renewable electricity and carbon capture from the atmosphere or waste CO_2 is called e-methanol (Verhelst et al. 2019).

Methanol, with the chemical formula CH_3OH, has a high hydrogen-to-carbon (H/C) ratio and does not form particulate matter since it is not a long-chain hydrocarbon (Verhelst et al. 2019). Considering its molar mass and lower heating value (LHV), methanol emits 20% less CO_2 compared to diesel, given similar engine efficiencies (Maritime Knowledge Center (MKC), TNO, TU Delft 2018). Another report indicates that methanol combustion in internal combustion engines results in approximately a 10% reduction in CO_2 emissions compared to heavy fuel oil (HFO) or distillate fuel (Gl 2019). The oxygen atom in the methanol molecule promotes more efficient combustion, requiring less air and resulting in lower CO_2 and PM emissions. Methanol also reduces soot emissions due to its oxygen content. Using methanol in diesel engines is a promising approach to reducing both soot and NO_X emissions (Tuner 2015; Zincir and Deniz 2021). The charge cooling effect of methanol, while it is injected into the cylinder, cools down the cylinder which results in lower NO_X emissions. Furthermore, the sulfur-free nature of methanol means no SO_X emissions from methanol combustion, enabling ships to comply with the IMO sulfur emission cap.

Methanol has low lubricity due to its lower kinematic viscosity compared to diesel. Therefore, lubrication additives are necessary when using methanol to prevent corrosion in injection pumps, injectors, and other fuel system components (Tuner 2015). Methanol is highly corrosive to certain materials. Its polar structure induces dry corrosion on these materials (Verhelst et al. 2019), affecting metals such as zinc, copper, lead, aluminum, and magnesium, as well as elastomers, plastics, and rubber (Methanol Institute 2016a, 2016b).

The IMO mandates that marine fuels must have a flashpoint higher than 60 °C for use on ships (International Maritime Organization (IMO) 1974). Methanol, however, is a low-flashpoint fuel with a flashpoint of 12 °C, necessitating additional safety precautions similar to those for LNG-fueled ships. Methanol's invisible flame and wide flammability limits (6.7–36%) require special attention. Despite these concerns, methanol vapor is heavier than air and can be extinguished with water (Aakko-Saksa et al. 2020).

Methanol is liquid at ambient temperature and pressure, similar to HFO, allowing for similar handling and safety practices with minor modifications to storage tanks

(Andersson and Salazar 2015; Stocker 2018; Zincir et al. 2019). Its high auto-ignition temperature reduces the risk of self-ignition or explosion. This results in lower capital costs compared to LNG storage. Methanol's infinite miscibility in water makes it less harmful to the environment and is suitable for double-bottom hull tank storage (Landälv 2017; Verhelst et al. 2019). However, its lower LHV means that roughly twice the tank volume of conventional fuel storage is required for the same navigation distance.

Environmentally, methanol is considered friendly due to its biodegradability and short half-life (1–6 days) in ground and water (Aakko-Saksa et al. 2020). Methanol dilutes quickly in water, breaking down into CO_2 and water, posing less ecological threat compared to other fuels, on the other hand, it increases sea vegetation (Center and (MKC), TNO, TU Delft 2018).

Methanol is toxic to humans, causing blindness if ingested in small amounts (10 mL) and potentially fatal at larger quantities (60–100 mL). It is also dangerous if absorbed through the skin or inhaled (Aakko-Saksa et al. 2020). Methanol is odorless below 2000 ppm in air, necessitating a completely closed fuel system, proper ventilation, and restricted contact on ships. Despite its toxicity, methanol is not carcinogenic, unlike diesel fuel (Andersson and Salazar 2015).

10.2.3 Biofuels

Biofuel is defined as the fuel obtained from the organic biomass of plants through various methods. Biofuels can be produced from a wide range of raw materials using different processes, resulting in end products with varying properties. There are three generations of biofuels: first, second, and third (Sevim and Zincir 2022).

First-generation biofuels are derived from edible food crops such as soybean, palm oil, rapeseed, sunflower, and sugarcane. These biofuels have been scaled up to commercial production and are now used in the transportation sector, including maritime transportation. Biodiesel (FAME) falls under the first generation, while renewable diesel (HVO) can be classified as both first-and second-generation biofuel depending on the raw materials used.

Second-generation biofuels are mainly produced from lignocellulosic materials and non-edible vegetable oils such as forest residues, jatropha, castor, and pongamia. They can also be derived from the same raw materials as first-generation biofuels. Examples of second-generation biofuels include FT-diesel (Fischer–Tropsch), pyrolysis oil, and HTL biocrude (hydrothermal liquefaction). The processing technologies for second-generation biofuels are more advanced due to the difficulty of breaking down the lignocellulose in plants. These processes produce a more stable fuel but require specific technologies that vary based on the biomass type, resulting in higher production costs (Kesharwani et al. 2019; Yuanrong Zhou 2020). First-generation biofuels are unlikely to meet all industry demands and be sustainable due to the large-scale use of farmland and competition with food. However, second-generation biofuels are not yet commercially viable because their production processes require expensive and advanced technology.

Given these limitations, researchers have focused on third-generation biofuel studies, particularly the production of biodiesel from microalgae. Microalgae biomass is considered a more sustainable biofuel source due to its advantages, such as not requiring quality farmland for production, yielding more biomass than conventional crops in the same area, having a high growth rate, and a short harvest cycle. Research into third-generation biofuels is still ongoing (Chowdhury and Loganathan 2019).

Biofuels are essentially combustible fuels created from biomass or bio-waste. There are many types of biofuels, distinguished by their production processes and raw materials. Among these, biodiesel, chemically known as Fatty Acid Methyl Ester (FAME), is the most mature in terms of both its production process and supply chain. In biodiesel production, fats and oils are converted into FAME through a process called esterification.

Another notable biofuel is renewable diesel, also known as Hydrotreated Vegetable Oil (HVO), due to its similarity with conventional diesel fuel (Kesharwani et al. 2019). HVO is produced from fats, vegetable oil, and used cooking oil (UCO) through a hydrotreating process known as fatty acids to hydrocarbon hydrotreatment. In this process, hydrogen is used to remove oxygen from triglyceride vegetable oil molecules, breaking the triglyceride into three chains. This results in hydrocarbons that are identical to conventional diesel fuel components, allowing HVO to be used in any ratio or in pure form without quality concerns in diesel engines (Engman et al. 2014).

HVO is highly compatible with both new and existing diesel engines and fuel systems. It meets the diesel fuel standards EN 590, except for having a lower density, and fully meets ASTM D975 standards (Engman et al. 2014). Biodiesel and HVO are renewable, clean-burning diesel replacement fuels that can be produced from a variety of raw materials, including plant oil, waste cooking oil, and animal fats.

10.2.4 Hydrogen

The International Energy Agency (IEA) recognizes hydrogen as a promising fuel to meet future energy demands (Qyyum et al. 2021) and anticipates its significant role in sustainable energy systems by 2050 (Staffell et al. 2019). Hydrogen, a carbon-free, and clean alternative fuel, produces only water as a by-product when combusted, making it an attractive option for maritime transportation.

Hydrogen's notable fuel properties include high laminar flame speed, wide flammability limits, high diffusivity, smaller quenching distance, and low minimum ignition energy (Deniz and Zincir 2016). It is a high-octane fuel, but its high auto-ignition temperature poses challenges for diesel engine combustion. However, hydrogen's low minimum ignition energy allows for quick combustion via external sources such as spark plugs. The wide flammability limits enable the use of variable fuel–air ratios, optimizing combustion conditions for lower fuel consumption and emissions. Hydrogen's high laminar flame speed enhances flame spread in the cylinder, contributing to more complete combustion, while high diffusivity ensures a more homogeneous fuel–air mixture, improving combustion efficiency.

Despite having a higher LHV than conventional marine fuels, hydrogen's energy density is low per volume. The volumetric energy capacity of gaseous hydrogen is 3000 times lower than diesel oil (Inal et al. 2021), necessitating six to seven times more storage space for the same energy content as HFO (Chryssakis et al. 2014). This low energy density is a significant drawback for hydrogen as a maritime fuel. Hydrogen can be stored as a compressed gas at 25 °C and 300 bar or as a cryogenic liquid at −253 °C and 1 bar (Inal et al. 2022).

Hydrogen is non-corrosive and non-toxic, reducing operational procedures and costs, although cryogenic storage is costlier than compressed storage due to the need for special insulation, cooling, and pumping systems to manage gasified hydrogen. Exposure to hydrogen concentrations over 336 mg/m^3 for 8 h can lead to asphyxiation (Inal et al. 2022).

Hydrogen's main advantages include its abundance, carbon-free nature, high energy conversion efficiency, multiple storage options, potential for long-distance transportation, higher LHV compared to conventional marine fuels, and ability to be converted into other energy forms (Dincer and Acar 2015).

10.2.5 Ammonia

Ammonia is a colorless gas with a sharp odor detectable at concentrations as low as 5 ppm in the air (Ayvalı et al. 2021). Composed of one nitrogen atom and three hydrogen atoms, ammonia is carbon-free and sulfur-free, resulting in combustion products that are free of carbonaceous emissions (CO_2, CO, unburned hydrocarbons), SO_X, and particulate matter (Zincir 2020). The combustion of ammonia primarily produces water, nitrogen, NO_X, and occasionally a small amount of nitrous oxide (N_2O).

Ammonia can be stored as a liquid at −33 °C and 1 bar or at room temperature and 10 bar (Ayvalı et al. 2021). Despite the ease of storing ammonia at room temperature and 10 bar, ammonia-fueled ships require 1.4–1.6 times more weight and 3.5–5.2 times more volume than HFO-fueled ships (Kim et al. 2020), and about 2.4 times larger tank volume compared to HFO (American Bureau of Shipping (ABS) 2020).

Ammonia is characterized by low reactivity, with a high auto-ignition temperature of 651 °C and narrow flammability limits of 16–25% by volume in air. Its burning velocity is 14 cm/s, which is lower than that of LNG (38 cm/s), LPG (40 cm/s), methanol (50 cm/s), and hydrogen (270 cm/s) (Zincir 2019). The low adiabatic flame temperature of ammonia (1800 °C) compared to methane (1950 °C), propane (2000 °C), and hydrogen (2110 °C) leads to unstable combustion at low engine loads and high engine speeds (Aziz et al. 2020). This instability arises because the in-cylinder temperature is insufficient to overcome high auto-ignition resistance at low loads and high engine speeds, and the combustion duration is not long enough for complete combustion due to high auto-ignition resistance and low flame velocity (Valera-Medina et al. 2018). Furthermore, the narrow flammability limits of ammonia can cause misfires in very lean or rich in-cylinder mixtures, thus restricting the operational range of internal combustion engines (Heywood 1988). Consequently, marine diesel engines are more suitable for ammonia combustion (with the assistance of pilot diesel injection) due to their large displacement volume, consistent high loads, and low engine speeds (Klüssmann et al. 2020).

Using ammonia as a fuel introduces several risks due to its properties. Ammonia is toxic, corrosive, and hazardous to aquatic life if spilled. Ammonia slip from the engine is another important issue. Therefore, ship crews must be thoroughly knowledgeable about these properties and implement additional precautions onboard (Zincir 2022).

Ammonia's toxicity is more than three times that of diesel, gasoline, and methanol (Aziz et al. 2020; McKinlay et al. 2020). Human contact with ammonia can result in three main types of injuries: dehydration, caustic burning, and freezing (Ayvalı et al. 2021). Dehydration occurs because ammonia extracts water from body tissues. Caustic burning results from ammonia's corrosive nature. Freezing happens as ammonia removes heat from the body during evaporation due to its high latent heat of vaporization. Even low concentrations of ammonia can affect the eyes, lungs, and skin, while exposure to high concentrations can be fatal (American Bureau of Shipping (ABS) 2020). Table 10.3 shows the risk levels and effects of ammonia exposure according to the acute exposure guideline levels (AEGL).

Another important property of ammonia is its corrosiveness. Ammonia is incompatible with certain materials, and the presence of moisture in the environment increases its chemical reactivity. The combustion of ammonia produces water, which can affect engine parts if incompatible materials are used. These materials include copper, brass, zinc, aluminum, and magnesium alloys (Klüssmann et al. 2020). Ammonia can change the color of incompatible surfaces to a greenish/blue hue and can lead to cracking (stress corrosion) in storage tanks and fuel supply lines made from steel, nickel steel, or carbon-manganese materials (American Bureau of Shipping (ABS) 2019). The risk of stress corrosion is heightened by oxygen contamination in liquid ammonia, necessitating the purging of air from tanks and fuel supply lines before ammonia bunkering (Brinks 2020). To prevent corrosion,

Table 10.3 Effect of ammonia exposure on humans (Jang et al. 2024)

Risk level	Exposure duration (min)			Exposure effect
	10	20	30	
AEGL-1	30 ppm	30 ppm	30 ppm	Irritation/discomfort/numb effect
AEGL-2	220 ppm	220 ppm	160 ppm	Irreversible-serious-long lasting adverse health effects
AEGL-3	2700 ppm	1600 ppm	1100 ppm	Life-threatening health effects/death

tanks, pipelines, and specific parts of the main and auxiliary engines should be made from steel, iron, and non-ferrous alloys (American Bureau of Shipping (ABS) 2020; Yapicioglu and Dincer 2019).

The toxicity of ammonia is well known, and a spill into the sea can pose a serious risk to aquatic life. Ammonia dissolves in water to form ammonium hydroxide (NH_4OH) and ammonium ions ($NH4^+$) (Ayvalı et al. 2021; Vries 2019), with undissolved ammonia evaporating into the atmosphere with a sharp odor. While ammonia is biodegradable in the long term, direct exposure is harmful and potentially fatal to organisms, though aquatic life can eventually recover through the nitrogen cycle.

Ammonia has a low fire risk compared to other potential alternative marine fuels. Its narrow flammability range (16–25% by volume in air), higher ignition energy requirement (two to three times that of conventional hydrocarbon fuels), and low burning speed (four times lower than methane) reduce the risk of fire on a ship (American Bureau of Shipping (ABS) 2020). However, ammonia's chemical reactivity with halogens and oxidizers can result in explosions. Therefore, ammonia must be stored away from these substances and ignition sources, with adequate ventilation provided.

Each alternative fuel has advantages and disadvantages. Table 10.4 shows the properties of alternative fuels, LNG, methanol, biodiesel, hydrogen, ammonia, and MDO. Ammonia comes forward with its carbon-free content and easy storage properties than other gaseous fuels, LNG, and hydrogen. Furthermore, ammonia has a similar volumetric density to diesel in liquid form (Karvounis et al. 2024). Another study states that ammonia is recognized as a prospective option for conventional fuels due to its high volumetric energy density and the possibility of production from renewable sources (Micco et al. 2024). From the lifecycle costs, ammonia has the lowest lifecycle cost among hydrogen and methanol (Zhou et al. 2024).

In this study, the system-based risk factors are identified according to the ammonia-related hazards. These hazards are listed in Table 10.5. Toxicity to humans and aquatic life is an important hazard type of ammonia. Corrosiveness and stress corrosion are the reason for the reactivity and material compatibility properties of ammonia. Low-temperature embrittlement occurs due to the lower storage temperature of ammonia in tank/tanks. A boiling liquid expanding vapor explosion (BLEVE) is the sudden vaporization of ammonia from the liquid state by external heating (fire) and violent explosion. Flammability is another ammonia fuel hazard, although flammability properties are lower than other alternative fuels.

10.3 Ammonia-Fueled Marine Diesel Engine System

This section contains the ammonia-fueled marine diesel engine system and potential hazardous critical equipment of the system. This study focuses on the ammonia high-pressure fuel supply system of the main propulsion engine.

Table 10.4 Fuel properties (Zincir 2022; Zincir and Deniz 2021; Sevim and Zincir 2022; Erdemir and Dincer 2020; Xing et al. 2021)

Properties	MDO	LNG	Methanol	Biodiesel	Hydrogen (gaseous/ liquid)	Ammonia
Carbon content (wt%)	84–87	75	50	77	–	–
Density (kg/ m^3)	796–841	430–470	792	880	17.5/71.1	602.8
Lower heating value (MJ/kg)	43.5	38.1	19.9	35	120.1	18.5
Energy density (MJ/ m^3)	36,403	22,020	15,600	30,800	2101/8539	11,333
Auto-ignition temperature (°C)	210	450–540	464	~210	571–585	651
Flashpoint (°C)	73.8	−184	12	100–170	–	-33
Boiling point (°C)	>180	−162	65	~ >180	−253	−33
Cetane/ Octane number	>35/–	–/130	<5 / 109	>47	–/>130	–/120
Flammability limits in air (vol%)	0.6–7.5	5–15	6.7–36	0.6–7.5	4–75	16–25
Storage state	Liquid	Cryogenic liquid	Liquid	Liquid	Compressed gas/ Cryogenic liquid	Compressed liquid/Cooled gas
Storage temperature (°C)	25	−162	25	25	25/–253	25/–33
Storage pressure (bar)	1	1	1	1	248/1	10/1

10.3.1 Bunkering Station

The bunkering station is the only area where the ammonia fuel system connects to the outside of the ammonia-fueled ship. The connection can be an ammonia fuel bunker ship or a shore-based bunkering connection. The main system equipment at the bunkering station is emergency release coupling (ERC) which is shown in Fig. 10.1. The ERC is a connection equipment on the ammonia bunkering line between the ship and outside of the ship. Its duty is to close and separate the connection when

Table 10.5 Ammonia fuel hazards (ClassNK 2022; European Maritime Safety Agency (EMSA) 2023)

Hazards	Consequence
Toxicity (humans)	Acute toxicity, skin burn, eye damage, damage to the respiratory system, frostbite (liquid form)
Toxicity (aquatic life)	Long-lasting toxic effect on aquatic life
Corrosiveness	Material degradation and equipment damage, incompatible with acids, brass, chlorine, copper, copper alloys, cadmium, mercury, peroxides, silver, and zinc
Stress corrosion	Stress corrosion cracking in carbon-manganese steel and nickel steel
Low-temperature embrittlement	Loss of material elasticity due to low temperature
BLEVE	The rupture of the tank results in explosion
Flammability	Low flammability between 15 and 28% by volume in air. But it can be burned and result in explosion. Pool fire (accumulation of combustible liquid) and flash fire (sudden fire for a short duration) are other hazards

there is an hazardous situation, such as excessive ship movement, misaligned connection, or damaged ERC (RINA 2024a). This area is occupied with a dry powder fire extinguishing system and a water curtain system.

10.3.2 Fuel Storage Area

The fuel storage area consists of ammonia fuel tank/tanks and related safety systems (pressure and temperature sensors, hydrocarbon sensors, and pressure relief valve). The fuel tank is the main fuel system equipment in the fuel storage area. There are four types of tanks for ammonia storage (Table 10.6).

The most common ammonia fuel tank is a Type C tank that can be used both pressurized and refrigerated. In this study, it is assumed that ammonia is stored at − 33 °C atmospheric pressure. The tank/tanks can be placed either under the deck or on the deck. Type C tank application to a ship is on deck application.

10.3.3 Fuel Preparation, Supply, and Recirculation

This section of the system, which is named the fuel preparation and supply system (FPSS), includes a buffer tank, high-pressure pumps, a vaporizer unit, and gas filters (Fig. 10.1). The FPSS is in a separate enclosed space fuel preparation room with a high-capacity ventilation system. The buffer tank maintains the ammonia fuel flow rate. High-pressure pumps increase ammonia fuel pressure to approximately 80 bar

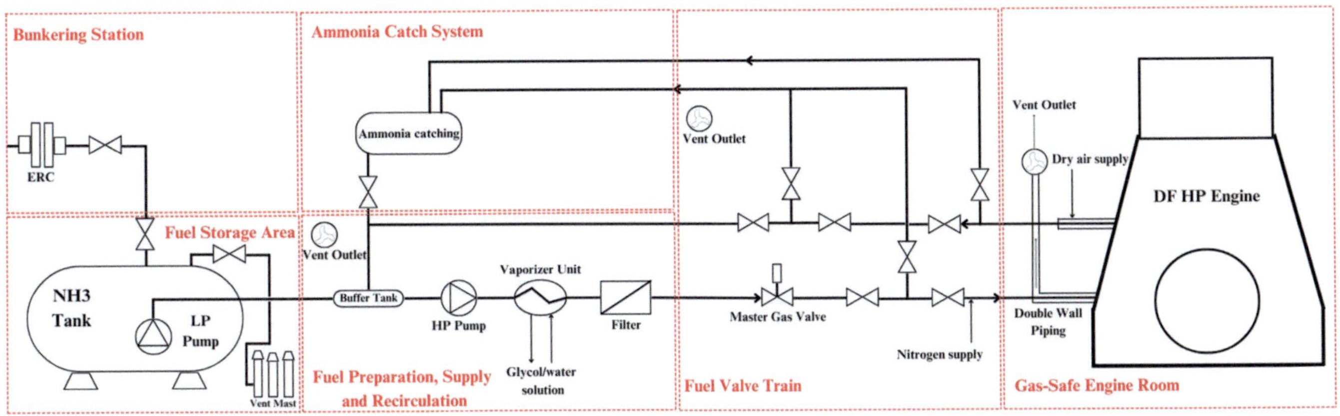

Fig. 10.1 High-pressure ammonia dual-fuel engine fuel preparation, supply and delivery system (figure combined, reproduced, and adapted) (ClassNK 2022; European Maritime Safety Agency (EMSA) 2023; MAN 2024)

Table 10.6 Ammonia fuel tank types (ClassNK 2022; International Maritime Organization (IMO) 2014)

Type	Type A	Type B	Type C	Membrane
Design pressure	$P_0 \leq 0.7$ bar	$P_0 \leq 0.7$ bar	$P_0 \geq 2$ bar	$P_0 \leq 0.7$ bar
Barrier type	Full secondary barrier	Partial secondary barrier	No secondary barrier	Full secondary barrier

for fuel supply (European Maritime Safety Agency (EMSA) 2023). The vaporizer unit uses a glycol-water solution to raise ammonia fuel temperature to 35–45 °C from −33 °C storage temperature (MAN 2017). The gas filter on the line prevents dirt and other contaminations move to the fuel valve train.

10.3.4 Fuel Valve Train

The fuel valve train is the safety valve group of the ammonia fuel system (European Maritime Safety Agency (EMSA) 2023). The fuel valve train is located in an enclosed space with a high-capacity ventilation system that is separated by the FPSS and engine room according to the International Code of Safety for Ship Using Gases and Other Low-Flashpoint Fuels (IGF Code) (International Maritime Organization (IMO) 2015). It is formed by a master gas valve, valve group on the line, pressure, temperature, and hydrocarbon sensors on the line. The master gas valve is the main safety valve of the fuel supply system. It cuts off the ammonia fuel supply if there is a leakage or any other unwanted situation that endangers the ship's safety. Remaining valve group is the line control valves that close or open the line after the fuel leak detection by the hydrocarbon sensors. In any case of emergency, 7–9 bar of nitrogen is delivered to the ammonia fuel line to purge ammonia from the line and main propulsion engine (MAN 2017).

10.3.5 Main Engine and Gas-Safe Engine Room

Gas-Safe Engine Room is one of the terms in the IGF Code (International Maritime Organization (IMO) 2015). This term means a ship engine room is protected from gas leakage (ammonia vapor) to the engine room and non-hazardous space is formed. Double-wall fuel supply piping with the ventilation connection is used in the engine room. If there is a leakage from the inner wall, ammonia vapor is quickly released into the atmosphere from the vent outlet (vent mast). Moreover, hydrocarbon sensors are placed on the line to detect the leakage. The main engine has also had a special design for preventing ammonia leakage. Double-wall piping with ventilation connection, hydrocarbon sensors, and sealing oil for preventing leakage from ammonia fuel injectors are the design items of an ammonia-fueled main propulsion engine.

Table 10.7 Dual-fuel engine specifications (ClassNK 2022; European Maritime Safety Agency (EMSA) 2023; MAN 2024)

	Low pressure (LP)	High pressure (HP)
Combustion cycle	Otto	Diesel
Combustion concept	Fuel nozzles on the cylinders inject low-pressure ammonia into the cylinders. Pre-mixed gas–air formation is burned by pilot diesel fuel	Injectors on the cylinder covers inject high-pressure ammonia into the cylinders after pilot diesel fuel is injected and burned. Diffusion combustion occurs
Fuel supply pressure	5–16 bar	80 bar
Injection pressure	5–16 bar	500–700 bar
Pilot fuel [%]	15–30	5–10
Minimum load for DF mode [%]	30	15
Combustion sensitivity	Fuel quality, fuel–air mixture ratio	N/A
Combustion issues	Ammonia slip, knock/misfire, reduced load response	Insignificant ammonia slip

Table 10.7 shows the specifications of low-pressure and high-pressure ammonia-fueled dual-fuel main engines. Despite two engine types being used for various alternative fuels (LNG, LPG, methanol), the combustion properties of ammonia are more deficient than other alternative fuels and high-pressure ammonia-fueled main engines are more suitable than the low-pressure one. In this study, it is assumed that the main engine is a high-pressure dual-fuel engine.

10.3.6 Ammonia Catch System

The duty of an ammonia catch system is to catch ammonia vapor and prevent the direct release of ammonia into the atmosphere when purging the system. After the system, there is only up to 5 ppm of ammonia vapor release into the atmosphere (MAN 2024).

10.3.7 Vent Mast

Vent masts are another hazardous line equipment. Fuel tanks, fuel preparation room, fuel valve train area, and main propulsion engine have ventilation connections. In an emergency situation, a high-capacity ventilation system sucks ammonia vapor into the atmosphere.

10.3.8 Critical Equipment and System-Based Risk Factors

The ammonia fuel supply and delivery system consists of various critical components and equipment. These are emergency release coupling (ERC), fuel tank, fuel preparation and supply system (FPSS), fuel valve train, double-wall piping, main engine, ammonia catch system, and vent mast.

To discuss the system-based risk factors, it is essential to focus on the area classifications that are explained in the IGF Code. Table 10.8 shows the area classifications and their definitions. According to the table, it can be seen that Area A class is the most hazardous area and Area D is the least hazardous area on an alternative-fueled ship. Identification of the critical equipment and system-based risk factors are done by considering the hazard possibilities of each equipment and the area where they are presented. Critical fuel system equipment, hazards, and area classifications are shown in Table 10.9. This table is formed by getting information from the literature on ammonia fuel safety and hazard studies (ClassNK 2022; European Maritime Safety Agency (EMSA) 2023; RINA 2024a; RINA 2024b).

According to the table, fuel tank is the most critical equipment in the ammonia fuel supply and delivery system, due to its seven types of hazards and area classification of A. Leakage from ammonia fuel tank affects ship crew and aquatic life by its toxicity. Furthermore, ammonia can react with tank material, and tank corrosion, stress corrosion, or low temperature embrittlement can occur. Although the flammability properties of ammonia are low, ammonia leakage can cause explosion and fire. Another important hazard is BLEVE, which arises when sudden vaporization and explosion of ammonia occurs in the tank due to quick heating of the tank by external factors. The second critical equipment is the ERC at the bunkering station. Despite it being a coupling between the bunkering station of the ship and the bunkering ship or shore-based supply facility, it is hazardous equipment in the system by six types of hazards and area classification of A. All hazards except BLEVE can happen when an unexpected event occurs. Excessive ship movement, misaligned connection, or disconnected coupling are some of the reasons for these hazards.

Table 10.8 Area classifications and definitions (ClassNK 2022; International Maritime Organization (IMO) 2015)

Area classification	Definition
A	Fuel always exists (Fuel tanks, piping). Hazardous area
B	Fuel does not exist normally, but it exists when a leakage (fuel preparation room, double-wall piping). Protected by double-wall piping system, ventilation system, and hydrocarbon sensors. Lower hazard potential than Area A
C	Fuel does not exist. Isolated area from Area A by two barriers and Area B by one barrier. Safe area
D	An enclosed space in Area C (accommodation spaces). No direct connection or contact to Area B

Table 10.9 Critical fuel system equipment, hazards, and area classifications

Equipment	Hazards	Area classification	Area specifications
ERC	Toxicity to humans Toxicity to aquatic life Corrosiveness Stress corrosion Low-temperature embrittlement Flammability	A	Open area Single-wall piping
Fuel tank	Toxicity to humans Toxicity to aquatic life Corrosiveness Stress corrosion Low-temperature embrittlement Flammability BLEVE	A	Open area Single-wall piping
FPSS	Toxicity to humans Corrosiveness Stress corrosion Low-temperature embrittlement Flammability	B	Enclosed area Single-wall piping High-capacity ventilation
Fuel valve train	Toxicity to humans Corrosiveness Stress corrosion Flammability	A	Enclosed area Single-wall piping High-capacity ventilation
Double-wall piping	Corrosiveness Stress corrosion	B	Enclosed area Double-wall piping High-capacity ventilation
Main engine	Corrosiveness Stress corrosion	B	Enclosed area Double-wall piping Regular ventilation Sealing oil
Ammonia catch system	Toxicity to humans Corrosiveness	A	Enclosed area Single-wall piping High-capacity ventilation
Vent mast	Toxicity to humans	A	Open area Single-wall piping

The least critical system component is the vent mast. It is because the only hazard is toxicity to the ship crew, but the probability of occurrence of this hazard is low. The ammonia catch system reacts with ammonia at the vent line and reduces the concentration below 5 ppm. Thus, the toxicity hazard can happen if the ammonia catch system does not work. The ammonia catch system is the second least critical system equipment on the system. Ammonia can be corrosive to the equipment. On the other hand, leakage from the equipment can cause toxicity hazards. However, the probability of these hazards is low, since compatible material is used in the system and a leak detection system is mounted on the equipment.

10.4 Summary

In this chapter, critical equipment and system-based risk factors of an ammonia-fueled marine diesel engine are identified. The chapter consists of the situation of maritime transportation, properties of LNG, methanol, biofuels, hydrogen, and ammonia, superiorities of ammonia over other alternative fuels, ammonia-fueled marine diesel engine system components, and critical equipment and system-based risk factors. The findings of the chapter are

- IMO has intensive GHG and CO_2 emission regulations. To comply with the new target of net-zero emissions by 2050, carbon-free clean-produced alternative fuels have to be used on ships.
- Ammonia is a carbon-free fuel with ease of storage. It has a similar volumetric density to diesel in liquid form. Ammonia is recognized as a prospective option to the conventional fuels due to its high volumetric energy density and possibility of production from renewable sources. From the lifecycle costs, ammonia has the lowest lifecycle cost among hydrogen and methanol. These properties make ammonia superior to other alternative fuels. To comply with net-zero emissions of IMO, ammonia can be produced by using renewable energy sources, such as solar energy or wind power.
- The ammonia fuel supply and delivery system consists of bunkering station, fuel storage area, fuel preparation and supply system, fuel valve train, main engine and gas-safe engine room, ammonia catch system, and vent mast. The ammonia fuel supply and delivery system is almost similar to other alternative fuel systems. The main difference is easy storage at −33 °C and 80 bar supply pressure which is lower than the LNG fuel system.
- The most critical equipment on the ammonia fuel supply and delivery system, due to its seven types of hazards and area classification of A. Leakage from ammonia fuel tank affects ship crew and aquatic life by its toxicity. Furthermore, ammonia can react with tank material, and tank corrosion, stress corrosion, or low temperature embrittlement can occur. Although the flammability properties of ammonia are low, ammonia leakage can cause explosion and fire. Another important hazard is BLEVE, which arises when sudden vaporization and explosion of ammonia in

the tank due to quick heating of the tank by external factors. The second critical equipment is the ERC at the bunkering station. Despite it being a coupling between the bunkering station of the ship and the bunkering ship or shore-based supply facility, it is hazardous equipment in the system by six types of hazards and area classification of A.

- The least critical system component is the vent mast. It is because the only hazard is toxicity to the ship crew, but the probability of occurrence of this hazard is low. The ammonia catch system reacts with ammonia at the vent line and reduces the concentration below 5 ppm. Thus, the toxicity hazard can happen if the ammonia catch system does not work. The ammonia catch system is the second least critical system equipment on the system. Ammonia can be corrosive to the equipment. On the other hand, leakage from the equipment can cause toxicity hazard. However, the probability of these hazards is low, since compatible material is used in the system and a leak detection system is mounted on the equipment.

Acknowledgements This work was supported by the Research Fund of Istanbul Technical University (Project ID: 45857). The authors declare that they have no known competing financial interests.

References

Aakko-Saksa PT, Westerholm M, Pettinen R, Söderström C, Roslund P, Piimakorpi P, Koponen P, Murtonen T, Niinistö M, Tuner M, Ellis J (2020) Renewable methanol with ignition improver additive for diesel engines. Energy Fuels 34:379–388

American Bureau of Shipping (ABS) (2015) LNG bunkering: technical and operational advisory. Technical Paper

American Bureau of Shipping (ABS) (2019) Setting the course to low carbon shipping. 2030 Outlook/2050 Vision

American Bureau of Shipping (ABS) (2020) Ammonia as marine fuel. Sustainability whitepaper, Oct 2020

American Bureau of Shipping (ABS) (2024) Beyond the Horizon. Carbon neutral fuel pathways and transformational technologies

Andersson K, Salazar CM (2015) Methanol as a marine fuel report. FCBI energy

Atchison J (2024) Enova funding for maritime ammonia: Amon Gas and Höegh Autoliners. https://ammoniaenergy.org/articles/enova-funding-for-maritime-ammonia-amon-gas-and-hoegh-autoliners/

Atchison J (2024) Ammonia marine fuel supply systems: Wartsila and Amogy. https://ammoniaenergy.org/articles/ammonia-marine-fuel-supply-systems-wartsila-and-amogy/

Ayvalı T, Tsang SCE, Van Vrijaldenhoven T (2021) The position of ammonia in decarbonising maritime industry: an overview and perspective: part II, costs, safety and environmental performance and the future prospects for ammonia in shipping. Johns Matthey Technol Rev 65(2):291–300. https://doi.org/10.1595/205651321X16127941688787

Aziz M, Wijayanta AT, Nandiyanto ABD (2020) Ammonia as effective hydrogen storage: a review on production, storage and utilization. Energies 13:3062. https://doi.org/10.3390/en13123062

Bilgili L (2021) Comparative assessment of alternative marine fuels in life cycle perspective. Renew Sustain Energy Rev 144:110985. https://doi.org/10.1016/J.RSER.2021.110985

Brinks HW, Hektor EA (2020) Ammonia as a marine fuel. DNV GL, Group Technology & Research, White Paper 2020
Chowdhury H, Loganathan B (2019) Third-generation biofuels from microalgae: a review. Curr Opin Green Sustain Chem 20:39–44. https://doi.org/10.1016/j.cogsc.2019.09.003
Chryssakis C, Balland O, Tvete HA, Brandsæter A (2014) DNV GL strategic research & innovation position paper 1–2014, alternative fuels for shipping
Class NK (2022) Part C "Guidelines for the safety of ships using ammonia as fuel" of guidelines for ships using alternative fuels. https://www.classnk.or.jp/hp/pdf/research/rd/2022/05_e07.pdf
ClassNK (2024) Press release. https://www.classnk.or.jp/hp/en/hp_news.aspx?id=11842&type=press_release&layout=1
de Vries N (2019) Safe and effective application of ammonia as a marine fuel. Delft University of Technology, Delft
Deniz C, Zincir B (2016) Environmental and economical assessment of alternative marine fuels. J Clean Prod 113:438–449. https://doi.org/10.1016/j.clepro.2015.11.089
Di Micco S, Cigolotti V, Mastropasqua L, Brouwer J, Minutillo M (2024) Ammonia-powered ships: concept design and feasibility assessment of powertrain systems for a sustainable approach in maritime industry. Energy Convers Manag: X 22:100539. https://doi.org/10.1016/j.ecmx.2024.100539
Dincer I, Acar C (2015) Review and evaluation of hydrogen production methods for better sustainability. Int J Hydrogen Energy 40:11094–11111. https://doi.org/10.1016/j.ijhydene.2014.12.035
DNV (2024) Energy transition outlook 2024. Maritime forecast to 2050. A deep dive into shipping's decarbonization journey
Elgohary MM, Seddiek IS, Salem AM (2014) Overview of alternative fuels with emphasis on the potential of liquefied natural gas as future marine fuel. Proc Inst Mech Eng Part M: J Eng MaritE Environ
Engman A, Hartikka T, Honkanen M, Kiiski U, Kuronen M, Mikkonen S, Saikkonen P (2014) Hydrotreated vegetable oil (HVO) premium renewable biofuel for diesel engines. Neste oil technical paper
Erdemir D, Dincer I (2020) A perspective on the use of ammonia as a clean fuel: challenges and solutions. Int J Energy Res 2020:1–8. https://doi.org/10.1002/er.6232
European Energy Agency (EEA) (2021) Emissions of air pollutants from transport. https://www.eea.europa.eu/data-and-maps/indicators/transport-emissions-of-air-pollutants-8/transport-emissions-of-air-pollutants-8
European Maritime Safety Agency (EMSA) (2023) Potential of ammonia as fuel in shipping International Maritime Organization (IMO) (2006) Code for the construction and equipment of ships carrying liquefied gases in bulk
Gl DNV (2019) Alternative fuels in the Arctic. Tech Rep 27(02):2019
Heywood JB (1988) Internal combustion engine fundamentals. McGraw-Hill, New York
Iannaccone T, Landucci G, Tugnoli A, Salzano E, Cozzani V (2020) Sustainability of cruise ship fuel systems: comparison among LNG and diesel technologies. J Clean Prod 260:121069. https://doi.org/10.1016/j.jclepro.2020.121069
Inal OB, Zincir B, Deniz C (2021) Comparison of hydrogen and ammonia for the decarbonization of shipping. In: 5th international hydrogen technologies congress (IHTEC-2021), May 26–28, 2021
Inal OB, Zincir B, Dere C (2022) Hydrogen as maritime transportation fuel: a pathway for decarbonization. In: Agarwal AK, Valera H (eds) Greener and scalable e-fuels for decarbonization of transport. Energy, environment, and sustainability. Springer, Singapore.
Independent Commodity Intelligence Services (ICIS) (2017) Chemical profile special. ICIS Chemical Business 2017 (26 May–1 June)
International Maritime Organization (IMO) (1974) International convention for the safety of life at sea (SOLAS)

International Maritime Organization (IMO) (2014) Resolution MSC.370(93) Amendments to the International Code for the Construction and Equipment of Ships Carrying Liquefied Gases in Bulk (IGC Code)

International Maritime Organization (IMO) (2015) Resolution MSC.391(95) Adoption of the international code of safety for ships using gases or other low-flashpoint fuels (IGF Code)

International Maritime Organization (IMO) (2020) Fourth IMO greenhouse gas study: executive summary

International Maritime Organization (IMO) (2022) Resolution MEPC.346(78) 2022 Guidelines for the development of a ship energy efficiency management plan (SEEMP)

International Maritime Organization (IMO) 2023 (2023) IMO strategy on reduction of GHG emissions from ships. https://www.imo.org/en/OurWork/Environment/Pages/2023-IMO-Strategy-on-Reduction-of-GHG-Emissions-from-Ships.aspx

Jang H, Mujeeb-Ahmed MP, Wang H, Park C, Hwang I, Jeong B, Zhou P, Papadakis A, Giannakis A, Sykaras K (2024) Safety evaluation on ammonia-fueled ship: gas dispersion analysis through vent mast. Int J Hydrogen Energy 83:1060–1077. https://doi.org/10.1016/j.ijhydene.2024.08.027

Karatuğ Ç, Ceylan BO, Ejder E, Arslanoğlu Y (2023) Investigation and examination of LNG, methanol, and ammonia usage on marine vessels. In: Zincir B, Shukla PC, Agarwal AK (eds) Decarbonization of maritime transport. Energy, environment, and sustainability. Springer, Singapore.

Karvounis P, Theotokatos G, Boulougouris E (2024) Environmental-economic sustainability of hydrogen and ammonia fuels for short sea shipping operations. Int J Hydrogen Energy 57:1070–1080. https://doi.org/10.1016/j.ijhydene.2024.01.058

Kesharwani R, Sun Z, Dagli C, Xiong H (2019) Moving second generation biofuel manufacturing forward: investigating economic viability and environmental sustainability considering two strategies for supply chain restructuring. Appl Energy 242:1467–1496. https://doi.org/10.1016/j.apenergy.2019.03.098

Kim K, Roh G, Kim W, Chun K (2020) A preliminary study on an alternative ship propulsion system fueled by ammonia: environmental and economic assessments. J Mar Sci Eng 8:183. https://doi.org/10.3390/jmse8030183

Klüssmann JN, Ekknud LR, Ivarsson A, Schramm J (2020) Ammonia application in CI engines. Special report, a report from the advanced motor fuels technology collaboration programme

Landälv I (2017) Methanol as a renewable fuel—a knowledge synthesis. Technical Report f3—The Swedish knowledge centre for renewable fuels

Lloyd's Register (LR) (2024) Ammonia-fuelled ore carrier design will be the first of this size in China. https://www.lr.org/en/knowledge/press-room/press-listing/press-release/2024/lr-awards-aip-to-maric-for-360000-dwt-ammonia-fuelled-vloc/

Lloyd's Register (LR) (2024) LR approval in principle for SDARI's 8,200 teu ammonia fuel boxship. https://www.lr.org/en/knowledge/press-room/press-listing/press-release/2024/lr-approval-in-principle-for-sdaris-8200-teu-ammonia-fuel-boxship/

Lloyd's Register (LR) (2024) LR grants AiP to SHI and Amogy for ammonia carrier powered by Amogy's ammonia-to-power technology. https://www.lr.org/en/knowledge/press-room/press-listing/press-release/2024/lr-grants-aip-to-shi-and-amogy-for-ammonia-carrier-powered-by-amogys-ammonia-to-power-technology/

Lloyd's Register (LR) (2024) Lloyd's Register grants AiP to HD KSOE for novel ammonia fuel supply system. https://www.lr.org/en/knowledge/press-room/press-listing/press-release/2024/lloyds-register-grants-aip-to-hd-ksoe-for-novel-ammonia-fuel-supply-system/

MAN (2017) Surveyors's day Gard/Norwegian Hull Club, MAN ME—dual fuel engines

MAN (2024) MAN B&W ammonia engine development. https://netl.doe.gov/sites/default/files/netl-file/ACT_0324_Wilche.pdf

Maritime Knowledge Center (MKC), TNO, TU Delft (2018) Methanol as an alternative fuel for vessels. Public Final Report, MIIP001-2017

McKinlay CJ, Turnock SR, Hudson DA (2020) A comparison of hydrogen and ammonia for future long distance shipping fuels. LNG/LPG and alternative fuels, 29–30 Jan 2020. The Royal Institution of Naval Architects, London

Methanol Institute (2016) Compatibility of elastomers in neat methanol service. Methanol Safe Handling Techn Bull

Methanol Institute (2016) Compatibility of metals & alloys in neat methanol service. Methanol Safe Handling Techn Bull

Nash, M (2015) IHS chemical: overview of the global methanol industry: the times they are achangin' IHS Chem

Qyyum MA, Dickson R, Shah SFA, Niaz H, Khan A, Liu JJ, Lee M (2021) Availability, versality, and viability of feedstocks for hydrogen production: product space perspective. Renew Sustain Energy Rev 145:110843. https://doi.org/10.1016/j.rser.2021.110843

RINA (2024a) Amendments to the IGF code and development of guidelines for alternative fuels and related technologies. Inherently safer ammonia-fuelled ship designs

RINA (2024b) Amendments to the IGF code and development of guidelines for alternative fuels and related technologies. Safe design of ammonia-fuelled vessels

Riviera (2024) Norway goes big on hydrogen and ammonia-fuelled ships. https://www.rivieramm.com/news-content-hub/news-content-hub/norway-goes-big-on-hydrogen--and-ammonia-fuelled-ships-81406

Rouwenhorst K (2024) Updated PGS-12 code: Preparing for increased ammonia imports to the Netherlands. https://ammoniaenergy.org/articles/updated-pgs-12-code-preparing-for-increased-ammonia-imports-to-the-netherlands/

Sevim C, Zincir B (2022) Biodiesel and renewable diesel as a drop-in fuel for decarbonized maritime transportation. In: Agarwal AK, Valera H (eds) Potential and challenges of low carbon fuels for sustainable transport. Energy, environment, and sustainability. Springer, Singapore

Staffell I, Scamman D, Velazquez Abad A, Balcombe P, Dodds PE, Ekins P, Shah N, Ward KR (2019) The role of hydrogen and fuel cells in the global energy system. Energy Environ Sci 12:463–491

Stocker A (2018) Investigation of renewable fuels for a Scottish ferry service. Thesis, Department of Mechanical and Aerospace Engineering, University of Strathclyde, Scotland, MSc

TradeWinds (2024) Eidesvik Offshore wins milestone approval for ammonia clean fuel system. https://www.tradewindsnews.com/offshore/eidesvik-offshore-wins-milestone-approval-for-ammonia-clean-fuel-system/2-1-1672767

Tuner M (2015) Combustion of alternative vehicle fuels in internal combustion engines. A report on engine performance from combustion of alternative fuels based on literature review. Report from a pre-study to prepare for interdisciplinary research on future alternative transportation fuels project

United Nations Conference on Trade and Development (UNCTAD) (2023) Review of Maritime Transport 2023

Valera-Medina A, Xiao H, Owen-JonesM DWIF, Bowen PJ (2018) Ammonia for power. Prog Energy Combust Sci 69:63–102. https://doi.org/10.1016/j.pecs.2018.07.001

Verhelst S, Turner JWG, Sileghem L, Vancoillie J (2019) Methanol as a fuel for internal combustion engines. Prog Energy Combust Sci 70:43–88

Xing H, Stuart C, Spence S, Chen H (2021) Alternative fuel options for low carbon maritime transportation: pathways to 2050. J Cleaner Prod 297:126651. https://doi.org/10.1016/j.clepro.2021.126651

Yao C, Pan W, Yao A (2017) Methanol fumigation in compression-ignition engines: a critical review of recent academic and technological developments. Fuel 209:713–732

Yapicioglu A, Dincer I (2019) A review on clean ammonia as a potential fuel for power generators. Renew Sustain Energy Rev 103:96–108. https://doi.org/10.1016/j.rser.2018.12.023

Yuanrong Zhou NP (2020) The potential of liquid biofuels in reducing ship emissions. International Council on Clean Transportation. https://theicct.org/publications/marine-biofuels-sept2020

Zhou X, Li T, Chen R, Wei Y, Wang X, Wang N, Li S, Kuang M, Yang W (2024) Ammonia marine engine design for enhanced efficiency and reduced greenhouse gas emissions. Nat Commun 15:2110. https://doi.org/10.1038/s41467-024-46452-z

Zincir B, Deniz C, Tuner M (2019) Investigation of environmental, operational and economic performance of methanol partially premixed combustion at slow speed operation of a marine engine. J Clean Prod 235:1006–1019

Zincir B (2019) An alternative fuel assessment model for ships and experiments on the effect of methanol on diesel engines. Dissertation thesis. Department of Maritime Transportation Engineering, Istanbul Technical University

Zincir B (2020) A short review of ammonia as an alternative marine fuel for decarbonised maritime transportation. In: Proceedings of international conference on energy, environment and storage of energy (ICEESEN2020), 19–21 Nov 2020, Kayseri-Turkey

Zincir B (2022) Ammonia for Decarbonized Maritime Transportation. In: Di Blasio G, Agarwal AK, Belgiorno G, Shukla PC (eds) Clean fuels for mobility. Energy, environment, and sustainability. Springer, Singapore

Zincir B, Deniz C (2021) Methanol as a fuel for marine diesel engines. In: Shukla PC, Belgiorno G, Di Blasio G, Agarwal AK (eds) Alcohol as an alternative fuel for internal combustion engines. Energy, environment, and sustainability. Springer, Singapore